Singularity Theory and Some Problems of Functional Analysis

Recent Titles in This Series

- 153 **S. G. Gindikin, Editor,** Singularity Theory and Some Problems of Functional Analysis
- 152 **H. Draškovičová, et al.,** Ordered Sets and Lattices II
- 151 **I. A. Aleksandrov, L. A. Bokut′, and Yu. G. Reshetnyak, Editors,** Second Siberian Winter School "Algebra and Analysis"
- 150 **S. G. Gindikin, Editor,** Spectral Theory of Operators
- 149 **V. S. Afraĭmovich, et al.,** Thirteen Papers in Algebra, Functional Analysis, Topology, and Probability, Translated from the Russian
- 148 **A. D. Aleksandrov, O. V. Belegradek, L. A. Bokut′, and Yu. L. Ershov, Editors,** First Siberian Winter School in Algebra and Analysis
- 147 **I. G. Bashmakova, et al.,** Nine Papers from the International Congress of Mathematicians 1986
- 146 **L. A. Aĭzenberg, et al.,** Fifteen Papers in Complex Analysis
- 145 **S. G. Dalalyan, et al.,** Eight Papers Translated from the Russian
- 144 **S. D. Berman, et al.,** Thirteen Papers Translated from the Russian
- 143 **V. A. Belonogov, et al.,** Eight Papers Translated from the Russian
- 142 **M. B. Abalovich, et al.,** Ten Papers Translated from the Russian
- 141 **Kh. Drashkovicheva, et al.,** Ordered Sets and Lattices
- 140 **V. I. Bernik, et al.,** Eleven Papers Translated from the Russian
- 139 **A. Ya. Aĭzenshtat, et al.,** Nineteen Papers on Algebraic Semigroups
- 138 **I. V. Kovalishina and V. P. Potapov,** Seven Papers Translated from the Russian
- 137 **V. I. Arnol′d, et al.,** Fourteen Papers Translated from the Russian
- 136 **L. A. Aksent′ev, et al.,** Fourteen Papers Translated from the Russian
- 135 **S. N. Artemov, et al.,** Six Papers in Logic
- 134 **A. Ya. Aĭzenshtat, et al.,** Fourteen Papers Translated from the Russian
- 133 **R. R. Suncheleev, et al.,** Thirteen Papers in Analysis
- 132 **I. G. Dmitriev, et al.,** Thirteen Papers in Algebra
- 131 **V. A. Zmorovich, et al.,** Ten Papers in Analysis
- 130 **M. M. Lavrent′ev, et al.,** One-dimensional Inverse Problems of Mathematical Physics
- 129 **S. Ya. Khavinson; translated by D. Khavinson,** Two Papers on Extremal Problems in Complex Analysis
- 128 **I. K. Zhuk, et al.,** Thirteen Papers in Algebra and Number Theory
- 127 **P. L. Shabalin, et al.,** Eleven Papers in Analysis
- 126 **S. A. Akhmedov, et al.,** Eleven Papers on Differential Equations
- 125 **D. V. Anosov, et al.,** Seven Papers in Applied Mathematics
- 124 **B. P. Allakhverdiev, et al.,** Fifteen Papers on Functional Analysis
- 123 **V. G. Maz′ya, et al.,** Elliptic Boundary Value Problems
- 122 **N. U. Arakelyan, et al.,** Ten Papers on Complex Analysis
- 121 **D. L. Johnson,** The Kourovka Notebook: Unsolved Problems in Group Theory
- 120 **M. G. Kreĭn and V. A. Jakubovič,** Four Papers on Ordinary Differential Equations
- 119 **V. A. Dem′janenko, et al.,** Twelve Papers in Algebra
- 118 **Ju. V. Egorov, et al.,** Sixteen Papers on Differential Equations
- 117 **S. V. Bočkarev, et al.,** Eight Lectures Delivered at the International Congress of Mathematicians in Helsinki, 1978
- 116 **A. G. Kušnirenko, A. B. Katok, and V. M. Alekseev,** Three Papers on Dynamical Systems
- 115 **I. S. Belov, et al.,** Twelve Papers in Analysis

(*Continued in the back of this publication*)

American Mathematical Society

TRANSLATIONS

Series 2 • Volume 153

Singularity Theory and Some Problems of Functional Analysis

S. G. Gindikin
Editor

American Mathematical Society
Providence, Rhode Island

Translation edited by SIMEON IVANOV

1991 *Mathematics Subject Classification.* Primary 40A05, 51F15, 57N75, 57R20, 57R45, 58C27, 58G28, 92C45; Secondary 12D99, 19E99, 28A99, 32G20, 35L99, 49N99, 60B99, 92E99.

Library of Congress Cataloging-in-Publication Data
Singularity theory and some problems of functional analysis/S. G. Gindikin, editor.
  p. cm. – (American Mathematical Society translations, ISSN 0065-9290; ser. 2, v. 153)
  Includes bibliographical references.
  ISBN 0-8218-7502-7 (alk. paper)
  1. Singularities (Mathematics) I. Gindikin, S. G. (Semen Grigor'evich) II. Series.
QA3.A572 Ser. 2, Vol. 153 QA614.58
510 s–dc20   92-32240
[514'.74]   CIP

COPYING AND REPRINTING. Individual readers of this publication, and nonprofit libraries acting for them, are permitted to make fair use of the material, such as to copy an article for use in teaching or research. Permission is granted to quote brief passages from this publication in reviews, provided the customary acknowledgment of the source is given.

Republication, systematic copying, or multiple reproduction of any material in this publication (including abstracts) is permitted only under license from the American Mathematical Society. Requests for such permission should be addressed to the Manager of Editorial Services, American Mathematical Society, P.O. Box 6248, Providence, Rhode Island 02940-6248.

The appearance of the code on the first page of an article in this book indicates the copyright owner's consent for copying beyond that permitted by Sections 107 or 108 of the U.S. Copyright Law, provided that the fee of $1.00 plus $.25 per page for each copy be paid directly to the Copyright Clearance Center, Inc., 27 Congress Street, Salem, Massachusetts 01970. This consent does not extend to other kinds of copying, such as copying for general distribution, for advertising or promotional purposes, for creating new collective works, or for resale.

Copyright ©1992 by the American Mathematical Society. All rights reserved.
Printed in the United States of America.
The American Mathematical Society retains all rights
except those granted to the United States Government.
The paper used in this book is acid-free and falls within the guidelines
established to ensure permanence and durability. ♾
This publication was typeset using $\mathcal{A}_{\mathcal{M}}\mathcal{S}$-TEX,
the American Mathematical Society's TEX macro system.
10 9 8 7 6 5 4 3 2 1   97 96 95 94 93 92

Contents

Foreword	vii
A. N. Varchenko, Period maps connected with a versal deformation of a critical point of a function, and the discriminant	1
V. A. Vassiliev, Characteristic classes of singularities	11
V. A. Vassiliev, Lacunas of hyperbolic partial differential operators and singularity theory	25
A. B. Givental', Reflection groups in singularity theory	39
V. M. Gol'dshteĭn and V. A. Sobolev, Qualitative analysis of singularly perturbed systems of chemical kinetics	73
V. V. Goryunov, Bifurcations with symmetries	93
S. M. Guseĭn-Zade, Stratifications of function space and algebraic K-theory	109
A. A. Davydov, Singularities in optimization problems	125
V. M. Zakalyukin, Nice dimensions and their generalizations in singularity theory	137
V. M. Klimkin, On an inverse problem of measure theory	149
S. Ya. Novikov, Classes of coefficients of convergent random series in spaces $L_{p,q}$	159
Yu. I. Sapronov, Corner singularities and multidimensional folds in nonlinear analysis	173
A. G. Khovanskiĭ, Newton polyhedra (algebra and geometry)	183

Foreword

The emergence of singularity theory marks the return of mathematics to the study of the simplest analytical objects: functions, graphs, curves, surfaces. Modern singularity theory for smooth mappings is one of the most intensely developing areas in mathematics which can be thought of as a crossroads where the paths joining the most abstract topics (algebraic and differential geometry and topology, complex analysis, invariant theory, Lie group theory) and the most applied topics (dynamical systems, equations of mathematical physics, asymptotics of integrals, geometrical optics, quasiclassical approximation in quantum mechanics, thermodynamics, mathematical economics, control theory) all meet each other.

Lectures in this volume include both reviews of the already established areas of singularity theory and the most recent results. In particular, in the lecture of A. N. Varchenko certain constructions are presented that are related to the discriminants of versal deformations of a critical point, and in the lecture by V. A. Vassiliev the qualitative behavior of a wave near its front is analysed. The description of singularities and their discriminants as orbit spaces for symmetry groups of appropriate right polyhedra is the subject of the lecture by A. B. Givental'. Corner singularities that appear in certain problems of nonlinear analysis are described, using the singularity theory approach, in the lecture by Yu. I. Sapronov. A. G. Khovanskiĭ demonstrates applications of algebraic geometry of polyhedra in algebra and in geometry.

An original approach to the analysis of critical phenomena in heat explosion theory is suggested in the lecture by V. M. Gol'dshteĭn and V. A. Sobolev, who applied methods of geometrical theory of singularly perturbed differential equations to problems of chemical kinetics.

All authors tried hard to reveal the meaningful part of the theory and not to conceal it behind formalism, notations, and technical details; attention is devoted to examples and discussions of results rather than to foundations and proofs. I hope that this collection helps to introduce to all mathematicians, as well as to physicists, engineers, and all other consumers of singularitytheory, the world of ideas and methods of this new area, and allows them to see how it can be applied to various fields and to learn how to use it themselves.

V. I. Arnol'd

Period Maps Connected with a Versal Deformation of a Critical Point of a Function, and the Discriminant

A. N. VARCHENKO

Let us consider a versal deformation of some singularity and the base of the versal deformation. To a general point of the base there corresponds an object in general position in the deformation. Singular fibers of the deformation determine a subset of the base, known as the discriminant. According to ideas going back to Thom [7], the pair (base of a versal deformation, the discriminant) conveys a significant amount of (if not all) information about the singularity. Thus, discriminants form an interesting class of hypersurfaces. The problem is to unravel the information hidden in them.

The following example illustrates the principle that the discriminant "remembers" the objects that are deformed over it.

Consider an algebraic function $z(\lambda_1, \ldots, \lambda_\mu)$, defined by an equation $P(z, \lambda) = 0$, where

$$P(z, \lambda) = z^N + a_1(\lambda_1, \ldots, \lambda_\mu) z^{N-1} + \cdots + a_N(\lambda_1, \ldots, \lambda_\mu).$$

Let $\Sigma \subset \mathbb{C}^\mu$ be its discriminant, i.e., the set of all λ for which the polynomial $P(z, \lambda)$ has multiple roots; let $\mathscr{L}_s : (\mathbb{C}^\mu, \Sigma) \to (\mathbb{C}^\mu, \Sigma)$ be an arbitrary continuous isotopy of the identity map. Then for any point λ, the number of distinct values of the function over the points of the curve $\mathscr{L}_s \lambda$ is constant (more precisely, the multiplicities of the roots of the polynomial $P(z, L_s \lambda)$ remain invariant under an isotopy of λ); see [3, 6, 17]. Thus, a change in the number of distinct values of an algebraic function is necessarily accompanied by a change in the discriminant topology.

In this report we give a number of constructions related to the discriminant of a versal deformation of a critical point of a function and to the period maps that arise in this situation; see [2, 5].

©1992 American Mathematical Society
0065-9290/92 $1.00 + $.25 per page

First, we make some general remarks explaining the elements of the constructions.

§1. The Gauss-Manin connection

Let $p : X \to \Lambda$ be a locally trivial bundle with the fiber X_λ. To this bundle we can associate the following locally trivial complex vector bundles of the fiber homology and cohomology:

$$p_k : \bigcup_{\lambda \in \Lambda} H_k(X_\lambda; \mathbb{C}) \to \Lambda,$$

$$p^k : \bigcup_{\lambda \in \Lambda} H^k(X_\lambda; \mathbb{C}) \to \Lambda, \qquad k = 0, 1, \ldots.$$

The bundles p_k and p^k are dual. The fibers of these bundles carry integral and real-valued structures—the images of the (co)homology with integer and real coefficients in the (co)homology with complex coefficients. Any cycle in a fiber can be continuously deformed to nearby fibers. The homology class of the deformed cycle does not depend on the deformation but is uniquely determined by the homology class of the original cycle. This construction determines connections in p_k and p^k, i.e., the possibility of parallel translations of points in a fiber over curves in the base. These connections are called the *Gauss-Manin connections*. Parallel translations along homotopic paths with fixed end points produce the same result. Therefore the connection has zero curvature. Parallel translation preserves integral and real structures. Translation over a closed path induces a linear automorphism of a distinguished fiber $H_k(X_\lambda; \mathbb{C})$. In this way we obtain a *monodromy representation* $\rho : \pi_1(\Lambda, \lambda) \to \operatorname{Aut} H_k(X_\lambda; \mathbb{C})$.

The image of the fundamental group is called the monodromy group of the bundle p_k. The definition of monodromy in p^k is similar. The Gauss-Manin connections in p_k and p^k are dual: the value of a covector at a vector is invariant under parallel translations.

If the base of the bundle is a complex manifold, then p_k and p^k are holomorphic vector bundles. Their holomorphic sections are defined as follows. If $e_1, \ldots, e_\mu$ is a local basis of continuous sections, taking values in the fiber (co)homology with integer coefficients, a holomorphic section is a section of the form $s = \sum s_j e_j$, where s_j are holomorphic functions on the base.

Let P be any section of p^k (or p_k). We define a multi-valued map $\tilde{P}$ of the base Λ into the distinguished fiber $H^k(X_{\lambda_0}; \mathbb{C})$: $\tilde{P}$ maps a point λ of the base to the value of the section $P(\lambda)$, translated to the distinguished fiber $H^k(X_{\lambda_0}; \mathbb{C})$ along any path in the base between points λ and λ_0. The result of the translation generally depends on the path connecting the points λ and λ_0. The branches of $\tilde{P}$ are obtained from one another by transformations from the monodromy group M of p^k. Hence $\tilde{P}$ determines a single-valued map $\overline{P}$ of the bundle base into the set of orbits $H^k(X_{\lambda_0};\mathbb{C})/M$ of the monodromy group on the fiber.

An isotopy of the identity map of the bundle's base is uniquely lifted by the Gauss-Manin connection to an isotopy of the spaces of the homology and cohomology bundles; in particular, it determines an isotopy of their sections. Two sections are said to be equivalent if there exists an isotopy of the base that transforms one section into the other.

Let Λ be a complex manifold and $\dim_\mathbb{C}\Lambda = \dim_\mathbb{C} H^k(X_\lambda;\mathbb{C})$. We say that a holomorphic section P of the bundle p^k is nondegenerate if the corresponding multi-valued map of the base into the distinguished fiber is a local diffeomorphism. We say that a section is stable if all nearby sections are equivalent to it.

PROPOSITION. *If Λ is a compact complex manifold, any nondegenerate section is stable.*

Indeed, let P_t be a deformation of a nondegenerate section P_0, and let $\tilde{P}_t$ be the corresponding deformation of the multi-valued map of the base into the distinguished fiber. Since $\tilde{P}_0$ is a local diffeomorphism, $\tilde{P}_t$ is also a local diffeomorphism for small t. We construct an isotopy h_t of the identity map of the base, which establishes an equivalence of sections. Such an isotopy not only exists, but is also unique and is determined by the following condition: h_t must carry a point λ of the base to a base point λ_t such that $\tilde{P}_t(\lambda_t) = \tilde{P}_0(\lambda)$. Since $\tilde{P}_0$ and $\tilde{P}_t$ are local diffeomorphisms, there is a unique point close to λ that satisfies this condition.

Using sections of the cohomology bundle, we can carry structures from the cohomology to the base. For example, let the fiber X_λ be a real $2k$-dimensional compact oriented manifold. Then $H^k(X_\lambda;\mathbb{C})$ contains a nondegenerate intersection form. If k is even, the intersection form is symmetric; if k is odd, it is antisymmetric. Let P be a nondegenerate holomorphic section of the bundle p^k. Then any branch of the multi-valued map $\tilde{P}$ identifies the tangent space at a point of the base with the tangent space at the corresponding point of the distinguished fiber of the cohomology bundle. This tangent space is, in turn, isomorphic to a fiber (as a tangent space to a linear space at a point). On the fiber we have a bilinear intersection form. Hence a multi-valued bilinear form is induced on the base. This form turns out to be single-valued, since the different branches of $\tilde{P}$ can be obtained from one another by monodromy transformations, and the intersection form is invariant with respect to the monodromy. The form on the base will be called the structure form and will be denoted by Φ.

The structure form is stable: if P_t is a small deformation of the initial section, there is an isotopy of the base that maps the structure forms Φ_t of the sections into one another.

PROPOSITION. *If k is odd, the structure form is a holomorphic symplectic structure on the base; if k is even, it is a holomorphic metric on the base with zero curvature.*

Indeed, in the first case the form Φ is induced by a holomorphic diffeomorphism from a nondegenerate differential 2-form with constant coefficients, and in the second case, from a metric with constant coefficients. Similarly, if k is odd and X_λ is a $2k$-dimensional (not necessarily compact) oriented manifold, then a nondegenerate holomorphic section of the k-dimensional cohomology bundle induces a holomorphic Poisson structure on the base, which is stable with respect to deformations of the section; see [2, 4, 5].

Sections of the cohomology bundle usually arise in the following way. Let ω be a differential k-form defined on the bundle space X and closed on each fiber. This form determines a section P_ω of the cohomology bundle p^k: to any point λ of the base the section assigns the cohomology class $[\omega|_{X_\lambda}] \in H^k(X_\lambda; \mathbb{C})$ of the restriction of the form to the fiber over the point. The section P_ω is called the period map of the form ω.

If X, Λ are holomorphic manifolds, $p : X \to \Lambda$ a holomorphic map, and ω a holomorphic differential form on X that is closed on fibers, then P_ω is a holomorphic section of the cohomology bundle. This follows from a standard analytical fact: if $\gamma_\lambda \subset X_\lambda$ is a k-dimensional cycle that depends continuously on a point λ over an open subset of the base, then the integral $I(\lambda) = \int_{\gamma_\lambda} \omega$ is a holomorphic function of λ.

Below we will discuss what these constructions give for the Milnor bundle of a versal deformation of a critical point.

§2. Milnor bundle of a critical point of a function [1, 2]

Let $f : (\mathbb{C}^n, 0) \to (\mathbb{C}, 0)$ be a germ of a holomorphic function with an isolated critical point. Consider a versal deformation of f (from which any deformation can be naturally induced). A versal deformation can be chosen in the form

$$F(x, \lambda) = f(x) + \lambda_1 \phi_1(x) + \cdots + \lambda_\mu \phi_\mu(x), \qquad (1)$$

where the ϕ_j generate a basis over $\mathbb{C}$ of the algebra

$$Q_f = \mathbb{C}(x)/(\partial f/\partial x_1, \ldots, \partial f/\partial x_n).$$

Choose $\phi_1(x) \equiv 1$. The dimension μ of the algebra Q_f over $\mathbb{C}$ is called the *Milnor number* of the germ f. Fix a sufficiently small ball B around the origin of the space $\mathbb{C}^n$, and then a sufficiently small (compared to B) ball Λ around the origin of the space $\mathbb{C}^\mu$. A point λ of Λ is called a *discriminant point* if the level manifold $X_\lambda = \{x \in B | F(x, \lambda) = 0\}$ is singular. The discriminant points form the discriminant Σ. The set of manifolds $\{X_\lambda\}$ forms a locally trivial bundle over the complement to the discriminant, $\Lambda \setminus \Sigma$, known as the Milnor bundle. A fiber of the Milnor bundle—the complex $(n-1)$-dimensional manifold X_λ—is homotopy equivalent to a bouquet of $(n-1)$-dimensional spheres. Denote the Milnor bundle by p. The associated bundles p^{n-1}, p_{n-1} with fibers $H^{n-1}(X_\lambda;\mathbb{C})$, $H_{n-1}(X_\lambda;\mathbb{C})$ are called the Milnor cohomology and homology bundles, respectively.

We can define an intersection form on $H_{n-1}(X_\lambda;\mathbb{C})$. It is symmetric if the number n of variables is odd, skew-symmetric if n is even. If the intersection form is nondegenerate, there is a dual nondegenerate intersection form on the dual space $H^{n-1}(X_\lambda;\mathbb{C})$. This is the case that will occupy us here.

If f is a germ of a function in three variables with a simple critical point of type A, D, E (see [2]), the intersection form on $H_2(X_\lambda;\mathbb{C})$ is negative definite and, in particular, nondegenerate. If f is a germ of a function in two variables, the intersection form on $H_1(X_\lambda;\mathbb{C})$ is nondegenerate if and only if the germ of the curve $f = 0$ is irreducible. There are simple formulas to check whether the intersection form is nondegenerate, in terms of the Newton polyhedron of the Taylor series of f, see [2, §14], and also [7, 16].

The bundles p_{n-1}, p^{n-1} possess many properties that were assumed in §1: the Milnor bundle is holomorphic; the dimension of the base $\Lambda\setminus\Sigma$ is equal to the dimension of a fiber of the Milnor cohomology bundle (and to the Milnor number); the index $n-1$ of the cohomology bundle under consideration is half the real dimension of the fiber X_λ.

§3. Period map in the Milnor bundle

Consider an arbitrary holomorphic differential $(n-1)$-form ω defined on $\mathbb{C}^n$. The restriction of ω to each fiber of the Milnor bundle is a closed form (since there are no nontrivial holomorphic n-forms on an $(n-1)$-dimensional holomorphic manifold). Let P_ω denote the period map of ω, that is, the section of the Milnor cohomology bundle that it generates. To express the period map in local coordinates, consider a basis $\gamma_1(\lambda),\ldots,\gamma_\mu(\lambda) \in H_{n-1}(X_\lambda;\mathbb{Z})$, defined for λ in some open subset of the base $\Lambda\setminus\Sigma$ and continuous in λ. This basis determines coordinates in the cohomology bundle fibers, in terms of which

$$P_\omega(\lambda) = \left(\int_{\gamma_1(\lambda)}\omega,\ldots,\int_{\gamma_\mu(\lambda)}\omega\right). \tag{2}$$

Together with the period map it is useful to consider the associated period map $P_\omega^{(l)}$, $l = 1, 2, \ldots$, that is, the section of the Milnor cohomology bundle defined by the formula

$$P_\omega^{(l)}(\lambda) = \left(\frac{\partial}{\partial\lambda_1}\right)^l\left(\int_{\gamma_1(\lambda)}\omega,\ldots,\int_{\gamma_\mu(\lambda)}\omega\right). \tag{3}$$

Recall that λ_1 is the free term in the formula (1).

EXAMPLE 1. Let $f = x_1^2 + \cdots + x_n^2$ be a germ of type A_1 at the origin. Then $F = f + \lambda_1$ is a versal deformation. The fiber of the Milnor homology bundle $H_{n-1}(X_{\lambda_1}; \mathbb{C})$ is one-dimensional and is generated by the vanishing cycle $\gamma(\lambda_1)$. The intersection form on $H_{n-1}(X_{\lambda_1}; \mathbb{C})$ is nondegenerate for odd n and vanishes for even n. The period map of a differential $(n-1)$-form is given by the integral of the form over the vanishing cycle:

$$P_\omega(\lambda_1) = \int_{\gamma_1(\lambda)} \omega = \operatorname{const} \lambda_1^{n/2} + \cdots.$$

Similarly,

$$P_\omega^{(l)}(\lambda_1) = \operatorname{const}(n/2)(n/2-1)\cdots(n/2-l)\lambda_1^{n/2-l} + \cdots.$$

We say that two holomorphic sections of the Milnor cohomology bundle are holomorphically equivalent if there exists an isotopy of the identity map of the pair (Λ, Σ), which consists of holomorphic diffeomorphisms, preserves the discriminant, and maps one section onto the other. We say that a holomorphic section is holomorphically stable if all nearby holomorphic sections are equivalent to it.

THEOREM 1 [5, 2]. (i) *If ω is a form in general position, the period map P_ω is nondegenerate in a suitable small neighborhood of the origin in Λ, minus the discriminant.*

(ii) *If the intersection form on $H_{n-1}(X_\lambda; \mathbb{C})$ is nondegenerate and ω is a form in general position, then for any $l \geq 0$ the associated period map $P_\omega^{(l)}$ is nondegenerate in a suitable small neighborhood of the origin in Λ, minus the discriminant.*

A form ω is in general position if its jet at $x = 0$, $\lambda = 0$ does not belong to a certain proper analytic subset of the jet space; for more details see [5, 1, 2]. The theorem states that the vector of integrals (2) (or (3)) for a form in general position defines a local (multi-valued) diffeomorphism of the complement of the discriminant to the affine cohomology space of the distinguished fiber of the Milnor bundle. In the above example, forms in general position are those with $\operatorname{const} \neq 0$. For example, $x_1 dx_2 \wedge \cdots \wedge dx_n$ is a form in general position.

THEOREM 2 [5, 2]. (i) *The period map of a differential form ω in general position is holomorphically stable. If the intersection form is nondegenerate, the associated period maps of forms ω in general position are holomorphically stable.*

(ii) *If the germ f is quasihomogeneous, the period maps of forms ω in general position are holomorphically equivalent in a neighborhood of the origin in Λ. Moreover, if the intersection form is nondegenerate, the period maps with fixed number l of forms ω in general position are holomorphically equivalent in a neighborhood of the origin in Λ.*

Recall that a germ is said to be *quasihomogeneous* if, for appropriate coordinates $x_1, \ldots, x_n$, weights $\alpha_1, \ldots, \alpha_n$, and all t,

$$f(t^{\alpha_1} x_1, \ldots, t^{\alpha_n} x_n) = t f(x_1, \ldots, x_n).$$

For example, simple germs A, D, E are quasihomogeneous. According to the theorem, for quasihomogeneous germs the period map does not depend on the choice of the form, up to a germ of a holomorphic diffeomorphism of the pair (Λ, Σ). In the above example the period map of a form in general position can be transformed to $P_\omega(\lambda_1) = \lambda_1^{n/2}$ by a change of the variable λ_1 that preserves the origin (discriminant). Similarly, the associated period map can be transformed to $P_\omega^{(l)}(\lambda_1) = \lambda_1^{n/2-l}$ if n is odd or $n/2 > l$.

Let V be the distinguished fiber of the Milnor cohomology bundle, M the monodromy group of the cohomology bundle and V/M its set of orbits. As noted in §1, for any section of the cohomology bundle and, in particular, for any associated period map $P_\omega^{(l)}$, we have an associated map $\overline{P_\omega^{(l)}}$ of the base $\Lambda \setminus \Sigma$ into the orbit set V/M. This map has a particularly simple structure for a germ f in an odd number of variables with a simple type A, D, E singularity. In this case the monodromy group is a finite group of the same type A, D, E, generated by reflections. A basis of polynomials invariant with respect to the monodromy determines an isomorphism $V/M \xrightarrow{\sim} \mathbb{C}^\mu$ and gives the orbit set the structure of a complex manifold. The corresponding map $\overline{P_\omega^{(l)}} : \Lambda \setminus \Sigma \to V/M$ is holomorphic. E. Looijenga [11] described a form ω in general position such that the map $\overline{P_\omega^{((n-1)/2)}} : \Lambda \setminus \Sigma \to V/M$ extends holomorphically to the discriminant and determines an isomorphism $\Lambda \xrightarrow{\sim} V/M$, which maps the discriminant onto the set of singular orbits. It follows from Theorem 2 that this assertion holds for all forms in general position. For generalizations of this construction to more complicated germs f in an odd number of variables, see the paper by A. B. Givental' in this volume. The case of germs of functions in an even number of variables remains open.

§4. Symplectic structure of the base of a versal deformation

In this section we assume that the intersection form on $H_{n-1}(X_\lambda; \mathbb{C})$ is nondegenerate. A nondegenerate dual form is then defined on the dual space $H^{n-1}(X_\lambda; \mathbb{C})$. The nondegenerate associated period map $P_\omega^{(l)}$ induces a bilinear form $\Phi_\omega^{(l)}$ on $\Lambda \setminus \Sigma$, which is symmetric for odd n and antisymmetric for even n.

THEOREM 3 [5, 2]. *Let $n = 2k$ be even. Then for any form ω in general position the structure form $\Phi_\omega^{(k-1)}$ can be extended holomorphically to the discriminant, giving a symplectic structure on the base Λ of a versal deformation.*

By Theorem 2, this symplectic structure is holomorphically stable with respect to deformations of the form ω; if the original germ is quasihomogeneous, the symplectic structure does not depend on the choice of ω. For example, if f is a germ of a function in two variables, then the form $\Phi_\omega = \Phi_\omega^{(0)}$ induced by the period map P_ω of a form ω in general position extends to a symplectic structure on the discriminant.

EXAMPLE 2. A_2: $F = y^2 + x^3 + \lambda_2 x + \lambda_1$, $\omega = y\,dx$, $\Phi_\omega = \operatorname{const} d\lambda_1 \wedge d\lambda_2$.

EXAMPLE 3. A_4:
$$F = y^2 + x^5 + \lambda_4 x^3 + \lambda_3 x^2 + \lambda_2 x + \lambda_1, \tag{4}$$
$\omega = y\,dx$, $\Phi_\omega = \operatorname{const}(d\lambda_4 \wedge d\lambda_1 + 3 d\lambda_3 \wedge d\lambda_2 + \lambda_4 d\lambda_4 \wedge d\lambda_3)$.

For other examples of symplectic and Poisson structures, see [2, 4]. If $n = 2k + 1$ is odd, the metric $\Phi_\omega^{(k)}$ corresponding to a form in general position has singularities on the discriminant, but off the discriminant it defines an isomorphism $T_*(\Lambda \setminus \Sigma) \to T^*(\Lambda \setminus \Sigma)$ of the tangent and cotangent bundles. It turns out that this isomorphism defines an isomorphism of the $\mathbb{C}(\lambda)$-modules of differential 1-forms on Λ with the vector fields tangent to the discriminant; see [5].

§5. Lagrangian properties of the discriminant

The discriminant splits in a natural way into strata, each stratum consisting of points λ over which the manifolds X_λ have the same degeneracy. For $n = 2k$, consider the symplectic structure $\Phi_\omega^{(k)}$ on the base. There is a principle according to which the degeneracy types of the manifolds X_λ must be reflected in the Lagrangian properties of the strata relative to this symplectic structure; see [5]. As an example, consider the stratum of the discriminant, which is the set of all λ such that the manifold X_λ has exactly $\mu/2$ singular points of type A_1.

THEOREM 4 [2, 5]. *This stratum is Lagrangian; that is, the restriction of the symplectic form to this stratum vanishes.*

More generally, let us restrict the symplectic structure on the tangent space to an arbitrary stratum at some point λ. We obtain an antisymmetric 2-form that is isomorphic to the intersection form on $H^{n-1}(X_\lambda; \mathbb{C})$ (note that X_λ is a singular manifold). For a rigorous account see [4].

Symplectic structure determines a field of characteristic directions on the discriminant, namely, the kernels of the restriction of the symplectic structure to the discriminant. Consider this field for the above example of an A_4 singularity. The points of the discriminant correspond to the curves X_λ defined by equation (4), in which the polynomial $x^5 + \lambda_4 x^3 + \cdots + \lambda_1$ has a multiple root. In this case X_λ is an elliptic curve with a point of self-intersection. The group $H_1(X_\lambda; \mathbb{Z})$ is the free abelian group of rank 3 generated by the basis cycles $\gamma_1(\lambda)$, $\gamma_2(\lambda)$ of the elliptic curve and by a "half-cycle" connecting the point of self-intersection with itself on the curve. It turns out that integral curves of the field of characteristics on the discriminant are defined by the equations $\int_{\gamma_j(\lambda)} y\,dx = \operatorname{const}_j$, $j = 1, 2$, i.e., they consist of parameters λ of elliptic curves with self-intersection, whose basic cycles $\gamma_1(\lambda)$, $\gamma_2(\lambda)$ have fixed areas. The singularities of the field of characteristics for A_4 are described in [4]; they arise from the singularities of the degeneracies of the symplectic structure on four-dimensional manifolds.

The question of what mutual properties of the discriminant and symplectic structure on the base of a versal deformation axiomatically determine the symplectic structure generated by the period map was not studied.

The period map of a form in general position induces a real and even an integral structure on the tangent bundle to $\Lambda \setminus \Sigma$. The question of what happens to these structures as one approaches the discriminant strata has not been studied. The answer to this question should probably be formulated in terms of mixed Hodge structure of singular manifolds lying over the strata.

For discriminants and period maps see also [9, 10, 12–15].

References

1. V. I. Arnol'd, A. N. Varchenko, and S. M. Guseĭn-Zade, *Singularities of differentiable mappings*, vol. I, "Nauka", Moscow, 1982; English transl., Birkhäuser, Basel, 1985.
2. _____, *Singularities of differentiable mappings*, vol. II, "Nauka", Moscow, 1984; English transl., Birkhäuser, Basel.
3. A. N. Varchenko, *Theorems of topological equisingularity of families of algebraic manifolds and polynomial maps*, Izv. Akad. Nauk SSSR Ser. Mat. **36** (1972), no. 5, 957–1019; English transl. in Math. USSR-Izv. **6** (1972), no. 5.
4. _____, *Period map and discriminant*, Mat. Sb. **134** (1987), no. 9, 66–71; English transl. in Math. USSR-Sb. **62** (1898), no. 1.
5. A. N. Varchenko and A. B. Givental', *Period map and intersection form*, Funktsional. Anal. i Prilozhen. **16** (1982), no. 1, 1–14; English transl. in Functional Anal. Appl. **16** (1982), no. 1.
6. A. N. Varchenko and S. M. Guseĭn-Zade, *Topology of caustics, wave fronts and degeneracies of critical points*, Uspekhi Mat. Nauk **39** (1984), no. 2, 190–191; English transl. in Russian Math. Surveys **39** (1984).
7. A. N. Varchenko and A. G. Khovanskiĭ, *Asymptotics of integrals over disappearing cycles and Newtonian polyhedron*, Dokl. Akad Nauk SSSR **283** (1985), no. 3, 521–525; English transl. in Soviet Math. Dokl. **32** (1985), no. 1.
8. R. Thom, *The bifurcation subset of a space of maps*, Manifolds, Lecture Notes in Math., vol. 197, Springer-Verlag, Berlin, pp. 202–208.
9. Ya. Karpishpan, *Hodge theory and hypersurface singularities*, Ph.D. Thesis, Columbia Univ., 1987.
10. _____, *Torelli theorems for singularities*, Invent. Math. **100** (1990), 97–141.
11. E. Looijenga, *A period map for certain semiuniversal deformations*, Compositio Math. **30** (1975), 299–316.
12. T. Oda, *K. Saito's period map for holomorphic functions with isolated critical points*, Algebraic Geometry (Sendai, 1985), Adv. Studies in Pure Math., vol. 10, North-Holland, Amsterdam and New York, 1987, pp. 591–648.
13. K. Saito, *Calcul algébrique de monodromie*, Asterisque **718** (1973), 195–211.
14. _____, *On the periods of primitive integrals*, Preprint, Harvard University, 1980.
15. _____, *Period map associated to a primitive form*, Publ. Res. Inst. Math. Sci. **19** (1983), 1231–1264.
16. _____, *Exponents and Newton polyhedra for isolated hypersurface singularities*, Math. Ann. **281** (1988), 411–417.
17. A. N. Varchenko, *Algebro-geometrical equisingularity and local topological classification of smooth maps*, Proc. International Cong. Math. (Vancouver, 1974), vol. 2, Canad. Math. Congress, Montreal, Que., 1975, pp. 427–431.

Translated by A. BOCHMAN

Characteristic Classes of Singularities

V. A. VASSILIEV

Let M, N be two smooth manifolds and $f: M \to N$ a smooth map. Consider the set of points in M where f has a singularity of a certain fixed class. If f is in general position and this class of singularities satisfies certain natural conditions, we can define a cohomology class of M dual to this set. Such classes are homotopic invariants of f; moreover, they can be used to construct bordism invariants of the manifold N. Many theorems on existence of singularities are proved with the help of characteristic classes dual to singularities; for example, if the class dual to singularities of f of type Σ is nontrivial, then any map homotopic to f has singularities of type Σ.

These invariant dual classes are expressed in terms of standard topological invariants of the triple (M, N, f), for example, in terms of characteristic classes of manifolds M, N and the action of f on them. The corresponding expressions are called the Thom polynomials of the singularities.

As a rule, singular sets of maps of real manifolds determine dual characteristic classes only in cohomology modulo 2. For a singularity to determine an integral characteristic class, it must be a cocycle of the universal complex of cooriented singularities, which will be described in §4. This complex is also used to prove numerous restrictions on the coexistence of singularities on one manifold, as well as to construct invariants of foliations.

Much topological information, too, may be obtained by studying the characteristic classes determined by sets of multiple points of maps (or, more generally, by cycles of multisingularities). In particular, these classes constitute obstructions to the embedding of smooth manifolds, while multisingularities can be used to prove additional conditions for the coexistence of singularities.

§1. Classification of singularities

Our lectures are structured in such a way that we must first recall some basic results about the classification of singularities of smooth maps. While

I shall discuss maps of real manifolds only, almost all the results and notions formulated below can be carried over to holomorphic maps of complex manifolds.

Let $f: M \to N$, $f': M' \to N'$ be smooth maps, $x \in M$, $x' \in M'$. The germs of f, f' at points x, x' are said to be RL-equivalent if the maps f, f' are locally defined by the same formulas for a suitable choice of coordinate systems near the points $x, x', f(x), f'(x')$. (In particular, the condition $\dim M = \dim M'$, $\dim N = \dim N'$ is necessary.) An RL-invariant classification of singularities is a classification such that RL-equivalent points are contained in the same class.

For example, by the implicit function theorem, all regular points of a map $M^m \to N^n$, that is, points at which the rank of the first differential is equal to $\min(m, n)$, are equivalent. Points at which this rank is less than $\min(m, n)$ are called *singular* points of the map and the number measuring the deficiency of the rank is the simplest invariant that distinguishes nonequivalent singular points (it is called the *corank* of the singularity).

DEFINITION. A point $x \in M$ belongs to the set $\Sigma^i(f)$ if the kernel of the operator df at x is i-dimensional.

If the set $\Sigma^i(f)$ is smooth, we can construct a second invariant—the rank of the restriction of f to the set $\Sigma^i(f)$ at a point x. As a rule, this can indeed be done, and for any nonincreasing finite sequence of nonnegative integers $I = (i_1, i_2, \ldots, i_k)$ and almost any map one can define the Thom-Boardman set $\Sigma^I(f)$. The definition is inductive and uses the following theorem. Let $I' = (i_1, \ldots, i_{k-1})$.

THEOREM ([7, 13]). *For almost any*([1]) *map f the set $\Sigma^{I'}(f)$ is a smooth submanifold of M.*

DEFINITION. Let the statement of the previous theorem hold for the sequence I' and map f. Then $\Sigma^I(f)$ is the set of points of the manifold $\Sigma^{I'}(f)$, at which the kernel of the differential of the restriction of f to this manifold is i_k-dimensional.

EXAMPLE. The two maps $\mathbb{R}^2 \to \mathbb{R}^2$ defined by the equations

$$\begin{cases} y_1 = x_1, \\ y_2 = x_1^2; \end{cases} \qquad \begin{cases} y_1 = x_1, \\ y_2 = x_2^3 - x_2 x_1; \end{cases}$$

are not equivalent: the first belongs to $\Sigma^{1,0}$ and the second to $\Sigma^{1,1,0}$. The number of nonzero entries i_l in the Thom-Boardman index I is called its *length* and is denoted by $\operatorname{ord} I$.

([1])Here and below, "almost any map" for M compact means "any map that belongs to some open dense subset of the space of maps"; and for arbitrary M this means "any map that belongs to a countable intersection of open dense subsets".

The Thom-Boardman classes can be defined for arbitrary (not necessarily "almost any") maps f as well. Consider the space $J_0^k(\mathbb{R}^m, \mathbb{R}^n)$ of all k-jets of maps $(\mathbb{R}^m, 0) \to (\mathbb{R}^n, 0)$, where $k = \operatorname{ord} I$. Boardman [7] defined a subset Σ^I of this space with the following important properties.

A. Σ^I is invariant under the obvious action on $J_0^k(\mathbb{R}^m, \mathbb{R}^n)$ of the groups of local (preserving 0) germs of diffeomorphisms of $\mathbb{R}^m$ and $\mathbb{R}^n$.

B. Σ^I is smooth and its closure is semialgebraic in $J_0^k(\mathbb{R}^m, \mathbb{R}^n)$.

Property A implies that for any pair of manifolds M, N the subset $\Sigma^I(M, N)$ of all jets in the space of k-jets $J^k(M, N)$ that are RL-equivalent to a jet in Σ^I is well defined. The obvious projection $J^k(M, N) \to M \times N$ turns this subset into the space of the locally trivial bundle with fiber Σ^I. For any smooth map $f: M \to N$ we can define the k-jet extension of f, $f_k: M \to J^k(M, N)$, which maps every point $x \in M$ to the k-jet of f at this point. The set $\Sigma^I(f)$ is defined as the preimage of $\Sigma^I(M, N)$ under this map. The words "almost any" may now be given the following rigorous meaning.

PROPOSITION. *For almost any map $f: M \to N$ and any k the maps f_k are transversal to all the manifolds $\Sigma^I(M, N)$.*

This is a corollary of Thom's transversality theorem (see [4, 15]). For the maps satisfying this condition the second definition of $\Sigma^I(f)$ coincides with the previous inductive one.

THEOREM ([7, 13, 4]). *Let $f: M^m \to N^n$ be a map in general position and I an arbitrary Thom-Boardman index. Then the codimension of $\Sigma^I(f)$ in M depends only on I and $m - n$ and is defined by the following formula:*

$$\nu_I(m-n) = (n-m+i_1)\mu(i_1, \ldots, i_k) - (i_1 - i_2)\mu(i_2, \ldots, i_k) - \cdots$$
$$- (i_{k-1} - i_k)\mu(i_k),$$

where $\mu(i_1, \ldots, i_k)$ is the number of integer sequences $j_1 \geq j_2 \geq \cdots \geq j_k$ such that $i_r \geq j_r \geq 0$ for every r $(1 \leq r \leq k)$, while $j_1 > 0$.

In particular, if $I = (i)$, then $\nu_i(m-n) = i(n-m+i)$; if $I = (i, j)$, then $\nu_I(m-n) = (n-m+i)i + j[(n-m+i)(2i-j+1) - 2i + 2j]/2$.

COROLLARY. *A nondegenerate map $f: M^2 \to N^2$ may only have singularities $\Sigma^{1,0}$ and $\Sigma^{1,1,0}$, but not $\Sigma^{1,1,1}$ and not $\Sigma^{\geq 2}$.*

REMARK. The Thom-Boardman classification is not complete. For example, for any $n \geq 2$ and sufficiently large k, the set

$$\Sigma^{2,0} \subset J_0^k(\mathbb{R}^n, \mathbb{R}^n)$$

contains three different orbits of the RL-group: two orbits of codimension 4 and an orbit of codimension 5 separating them; these subclasses of the class $\Sigma^{2,0}$ are called the *elliptic, hyperbolic,* and *parabolic* modifications, respectively, of the singularity $\Sigma^{2,0}$ (see [4]).

§2. Cohomology classes dual to singularities

Let S be any semialgebraic RL-invariant subset of codimension ν in the space $J_0^k(\mathbb{R}^m, \mathbb{R}^n)$ and $S(M, N)$ the corresponding subset of the space $J^k(M, N)$. (For example, $S(M, N)$ may be the closure of one of the classes Σ^I). Then, according to [8] we can define a class $[S] \in H^\nu(J^k(M, N), \mathbb{Z}_2)$ which is the Poincaré dual of $S(M, N)$. Moreover, for nondegenerate maps $M \to N$ the analogous cohomology class of M can be defined. To formulate the result, let us choose an arbitrary Whitney stratification of the set S (see [15]).

THEOREM ([20, 11]). *If the map $f : M \to N$ is such that the corresponding jet extension $f_k : M \to J^k(M, N)$ is transversal to the stratified set S (this is true for almost every map f), then the set $S(f) = f_k^{-1}(S)$ defines an element in the homology group $H_{m-\nu}^f(M, \mathbb{Z}_2)$ of the manifold M with closed support and, consequently, defines a dual class $[S(f)]$ in the group $H^\nu(M, \mathbb{Z}_2)$. Moreover, the latter class coincides with the class $f_k^*([S])$.*

COROLLARY. *The class $[S(f)]$ is invariant under smooth homotopies of f. Indeed, a homotopy of maps defines a homotopy of their jet extensions.*

Moreover, now we can define the dual class of $S(f)$ even when the map f is not in general position (when the set $S(f)$ may have arbitrary irregular structure): by definition, this class is always $f_k^*([S])$.

The homotopic invariance of the classes $[S(f)]$ implies numerous results concerning the existence of singularities: if such a class is nontrivial then any map homotopic to f must have singularities of type S. For examples of such theorems, see the next sections.

§3. Thom polynomials

Recall that for any m-dimensional manifold M we can define m cohomology classes $w_i(M) \in H^i(M, \mathbb{Z}_2)$, $i = 1, \ldots, m$; w_i are the Stiefel-Whitney classes of the tangent bundle of M (in the case of complex manifolds the corresponding objects are the Chern classes $c_i(M) \in H^{2i}(M, \mathbb{Z})$, see [14]). Let $f : M^m \to N^n$ be a smooth map. Then it is possible to define n more classes on M, $w'_i \equiv f^*(w_i(N)) \in H^i(M, \mathbb{Z}_2)$.

THOM'S THEOREM (see [11]). *Let S be an arbitrary analytic RL-invariant subset of the space $J_0^k(\mathbb{R}^m, \mathbb{R}^n)$. Then there exists a universal polynomial T_S over $\mathbb{Z}_2$ in $m + n$ variables, depending only on S, m, and n, such that for any smooth manifolds M^m, N^n and any map $f : M^m \to N^n$ the class*

$[S(f)] \in H^*(M^m, \mathbb{Z}_2)$ *dual to the singular set* $S(f)$ *is equal to*

$$T_s(w_1(M), \ldots, w_m(M), w_1'(N), \ldots, w_n'(N)).$$

As a rule, this statement is used when S is the closure of one of the Thom-Boardman classes. The polynomial T_S is called the *Thom polynomial* (T.p.) for the set S.

EXAMPLE. Let $m = n$, and let $S = \overline{\Sigma}^1$ be the set of all singular points of a map. Then for any f the class dual to $S(f)$ is the difference of the first Stiefel-Whitney classes of the tangent bundles TM and f^*TN: $[S(f)] = w_1(TM) - w_1'(TN)$. If N is orientable, this class is responsible for the orientability of the manifold M: it is orientable if and only if $S(f)$ is homologous to zero in M.

The present situation with the computation of T.p. for the classes Σ^I is as follows. T.p. have been completely computed for the first-order singularities ($\Sigma^I = \Sigma^i$); see [16]. In the case $I = (i, j)$ an algorithm computing T.p. is given in [16, 17], and the computation is actually carried out in [17] for a few examples. For more complicated singularities there are only a few disconnected results. In all these cases the Thom polynomials express $[S(f)]$ in terms of not all the classes $w_i(M), w_j'(N)$, but only of the ratio of the total Stiefel-Whitney classes of TM and f^*TN. Let us recall this notion.

The *total Stiefel-Whitney class* of an l-dimensional vector bundle $L \to M$ is the element

$$1 + w_1(L) + \cdots + w_l(L) \in H^*(M, \mathbb{Z}_2),$$

denoted by $w(L)$. Let $A(M)$ be the subset of $H^*(M, \mathbb{Z}_2)$, consisting of all elements $a = a_0 + a_1 + \cdots$, $a_i \in H^i(M, \mathbb{Z}_2)$, such that a_0 is the unit element of the ring $H^*(M, \mathbb{Z}_2)$. $A(M)$ is a multiplicative abelian group and contains the total Stiefel-Whitney classes of all vector bundles over M, products and quotients of these classes, etc.

NOTATION. Let $f: M \to N$ be a smooth map. Then $w(f)$ denotes the element $w(TM)/f^*w(TN)$ of $A(M)$, that is, the ratio of the total Stiefel-Whitney class of TM and the f-preimage of the total class of TN.

Thom polynomials for the first-order singularities Σ^i. Let $\overline{w}_l$ denote the lth homogeneous component of the class $w(f)$, so that $w(f) = 1 + \overline{w}_1 + \overline{w}_2 + \cdots$. Let $j = \min(m, n) - i$.

THEOREM (see [16]). *For any* $i = 1, 2, \ldots$ *the class*

$$[\Sigma^i(f)] \in H^{i(n-m+i)}(M, \mathbb{Z}_2)$$

is equal to the determinant of the following matrix:

$$\begin{pmatrix} \overline{w}_{m-j} & \overline{w}_{m-j+1} & \cdots & \overline{w}_{m+n-2j-1} \\ \overline{w}_{m-j-1} & \overline{w}_{m-j} & \cdots & \overline{w}_{m+n-2j-2} \\ \cdots & \cdots & \cdots & \cdots \\ \overline{w}_{m-n+1} & \overline{w}_{m-n+2} & \cdots & \overline{w}_{m-j} \end{pmatrix}. \qquad (1)$$

EXAMPLE. Let $m < n$, $i = 1$, i.e., $j = m - 1$. Replace $\overline{w}$ in (1) by the usual Stiefel-Whitney classes of the tangent bundle of M. Suppose that the determinant (1) thus obtained does not vanish as a class in $H^{n-m+1}(M, \mathbb{Z}_2)$. Then M cannot be immersed in $\mathbb{R}^n$, since the set of singularities Σ^1 of any map $M \to \mathbb{R}^n$ is not empty. (This is not surprising, since when $j = m - 1$ the determinant (1) of any element $a = 1 + a_1 + \cdots \in A(M)$ is exactly equal to the $(n - m + 1)$-dimensional component of the element $a^{-1} \in A(M)$, i.e., in our case, to the $(n - m + 1)$-dimensional Stiefel-Whitney class of the normal bundle of M for any immersion of M in Euclidean space.)

§4. Integral characteristic classes and universal complexes of singularities

In general, the Thom-Boardman class (or another class of singularities) of a real manifold defines a dual element only in the cohomology with coefficients in $\mathbb{Z}_2$, but not in $\mathbb{Z}$. For example, the singular set of a typical map $\mathbb{R}P^2 \to \mathbb{R}^2$ does not define any element of $H^1(\mathbb{R}P^2, \mathbb{Z})$: this is because there is no invariant way of assigning signs "+" or "−" to the points of the transversal intersection of this set with oriented curves in $\mathbb{R}P^2$. To define a dual class in the integral cohomology, the class of singularities must be a cocycle of a certain universal complex (see [21, 22]); the generators of this complex are the cooriented singularity classes, with the differentials defined by the adjacency relations among classes of neighboring codimensions. Every "singularity theory", of smooth functions, Lagrange and Legendre maps, general maps of smooth manifolds, etc., has its own universal complex, which conveys a variety of topological information about the global properties of the singularities.

EXAMPLES. *Maslov index and first Pontryagin class.* Let N be a Lagrange submanifold immersed in the tangent bundle T^*W of a manifold W (see [12, 4]). If this immersion $N \hookrightarrow T^*W$ is in general position, the singular set of the projection $N \to W$ has codimension 1 in N. It turns out that the index of the intersection with this set already defines a class in the integral cohomology of N (see [12, 1]). (It follows, in particular, that $\mathbb{R}P^2$, like any other surface with odd Euler characteristic, does not have a Lagrange immersion in $T^*\mathbb{R}^2$.) This is due to two important properties of the simplest Lagrange singularity (A_2-fold)—its coorientability and the regular behavior of the coorientation near the more complicated singularities.([2]) These properties are as follows.

([2])Recall (see, for example, [4]) that the classification of singularities of a Lagrange projection is determined by the classification of the corresponding singularities of the generating functions, modulo a stable equivalence. In particular, the most widespread Lagrange singularities, i.e., those with the least codimension, are

$$A_k^\pm (k \geq 2),\ D_k^\pm (k \geq 4),\ E_6,\ E_7,\ E_8,\ P_8^{1,2},$$

etc.; see [4, 3].

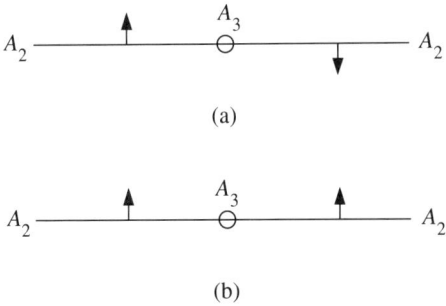

FIGURE 1

Since the immersion $N \to T^*W$ is in general position, the singular set is the closure of the A_2-type points; it contains the set of A_3-type points as a smooth subset of codimension 1, and the set of singularities of other types as a subset of codimension 2 or more. It turns out that in the neighborhood of any A_2-type point the set $\{A_2\}$ locally divides N into two inequivalent components: there is an invariant way to call one of them positive and the other negative (see [1]). A typical curve in N has only transversal intersections with the singular set at the points of $\{A_2\}$. The Maslov index of an oriented closed curve is defined as the number of times it crosses from the negative side to the positive side, minus the number of crossings in the opposite direction. It turns out that this number depends only on the class of the curve in $H_1(N, \mathbb{Z})$. This would be impossible if the transversal orientations of the components of A_2 were poorly compatible near the A_3-points, for example, as shown by the arrows in Figure 1a. Indeed, in that case the Maslov index of the small circle with center at A_3 would be 2, but the circle is homologous to zero.

In reality, however, the coorientations of A_2 are compatible as shown in Figure 1b, so that it is possible to define the Maslov index as an integral cohomology class. The condition that the components of the singular set have compatible orientation can be formalized and generalized to the more complicated singularities as the property of being a cocycle of the universal complex of singularities. For example, in this complex the pattern of Figure 1a would correspond to the formula $\delta(A_2) = \pm 2A_3$, whereas the right situation would correspond to $\delta(A_2) = 0$.

A second important example deals with the theory of singularities of general (non-Lagrange) maps. Let M^n be a compact manifold and $f: M^n \to \mathbb{R}^n$ a map in general position.

PROPOSITION. *The sets* $\Sigma_e^{2,0}$ *and* $\Sigma_h^{2,0}$ *of elliptic and hyperbolic singular points of type* $\Sigma^{2,0}$ *(see end of* §1*) have an invariant coorientation in* M^n *(that is, the orientation of any transversal germ of a four-dimensional submanifold of* M^n*).*

It is enough to check the compatibility of these coorientations in the neighborhood of the simplest degenerate points of $\Sigma^{2,0}$ (that is, the parabolic points $\Sigma_p^{2,0}$). However, near these points each of the classes $\Sigma_e^{2,0}$, $\Sigma_h^{2,0}$ is represented by one component, so we can choose their standard coorientations to be compatible. Then the index of intersection with the cooriented set $\Sigma^{2,0}$ will give us a class $[\Sigma^{2,0}] \in H^4(M, \mathbb{Z})$.

THEOREM (see [20]). *If the compatible coorientations of the sets $\Sigma_{e,h}^{2,0}$ are suitably chosen, the class $[\Sigma^{2,0}]$ coincides, modulo torsion, with the first Pontryagin class of the manifold M.*

Universal complexes of singularities are differential complexes, whose formal generators are invariantly coorientable classes of singularities (that is, classes, such that, for any nondegenerate map, the manifold of singular points of the corresponding type has a standard, invariantly defined orientation of the normal bundle in the preimage manifold). A change of coorientation corresponds to multiplication of a generator by -1, the dimension of the generator is equal to the codimension of the class of singularities, and differentials (incidence coefficients) are defined naturally by the behavior of the components of the classes of lower codimension near the points of classes of higher codimension. The cohomology of these complexes is invariant to the homotopies of the map. Moreover, it defines invariants of the corresponding oriented bordisms (see [5, 22]).

The cohomology of complexes of singularities of Lagrange maps is computed in [21, 22] up to dimension 6 inclusive. Apart from the one-dimensional Maslov class dual to the (suitably cooriented) singularity set A_2, there is also a five-dimensional class, dual to the set of singularities A_6 (or the homological set E_6) and a six-dimensional class dual to the set of singularities P_8. The value of the corresponding characteristic classes on the fundamental cycle of an oriented compact five-dimensional (respectively, six-dimensional) Lagrange manifold is an invariant of an oriented Lagrange cobordism (see [2]) and these invariants can be realized (see [9, 6]).

Another application of universal complexes is due to the fact that they provide numerous restrictions on the coexistence of singularities on a single manifold: if some class (or a linear combination of classes) is a coboundary of a complex, the corresponding cycle in the manifold is homologous to zero. For example, if $\delta(A_5) = E_6 - A_6$, the numbers of points of types A_6 and E_6, taken with their natural signs, are equal on any five-dimensional Lagrange manifold (a point of type A_6 or E_6 is taken with the " + " or " − " according to whether or not the orientation of the manifold at the point is equal to the coorientation of the singularity). Restrictions on the parity of the number of the singular points are particularly easy to formulate; for example, on typical compact (not necessarily oriented) Lagrange manifolds of suitable dimension n, the numbers of the following singularities are even: $A_4 + D_4$ for $n = 3$,

D_5 for $n = 4$, D_6 and $A_6 + E_6$ for $n = 5$, D_7 and $E_7 + P_8$ for $n = 6$, and $A_8 + E_8$ and P_9 for $n = 7$. These results follow from computations in the universal complex of noncooriented singularities (this complex is the space over $\mathbb{Z}_2$ generated by all, not only cooriented, classes of singularities; it provides invariants of a nonoriented Lagrange cobordism (see [2])).

Similar results must hold, of course, for other singularity theories, in particular, for singularities of general (non-Lagrange) maps of manifolds; a special attention is being paid here to the Lagrange case, since it is the most investigated one.

We note one more application of universal complexes: cohomologies of complexes of singularities of functions (or, more generally, maps to $\mathbb{R}^p$) are invariants of smooth bundles and even foliations (see [5]).

§5. Multiple points and multisingularities

Formula for multiple points of immersions and obstructions to embedding of manifolds. Let $m < n$ and let f be a smooth map of an m-dimensional compact manifold M into an n-manifold N. Let N_k be the set of multiplicity k points of the set $f(M)$, i.e., points $y \in N$ such that $\#f^{-1}(y) = k$; let $M_k = f^{-1}(N_k)$. If f is an immersion in general position, the closures of M_k and N_k are cycles of dimension $n - (n-m)k$ in M and N. Denote the dual elements of $H^*(M, \mathbb{Z}_2)$, $H^*(N, \mathbb{Z}_2)$ by m_k, n_k. These elements are not changed by deformations of the immersion f. Let $e \in H^{n-m}(M, \mathbb{Z}_2)$ be the $(n-m)$th Stiefel-Whitney class of the normal bundle $\nu = f^*TN/TM$ on M.

THEOREM (see [18]). *Let f be an immersion in general position. Then*

$$m_k = f^*(n_{k-1}) - e \smile m_{k-1}. \qquad (2)$$

If, in addition, the number $n - m$ is even and the bundle ν is orientable (so that its Euler class $e \in H^{n-m}(M, \mathbb{Z})$ is defined, see [14]), then all the m_k, n_k may also be defined as integral cohomology classes and formula (2) *is true in $H^*(M, \mathbb{Z})$ as well.*

EXAMPLE. For any immersion of M into Euclidean space, $w(TM) \smile w(\nu) = 1$, that is, $w(\nu) = (w(TM))^{-1} \in A(M)$. It follows that if the $(n-m)$-dimensional homogeneous component of $(w(TM))^{-1}$ does not vanish, M cannot be embedded in $\mathbb{R}^n$. Indeed, for any immersion $M \to \mathbb{R}^n$, when $k = 2$ in formula (2), the class n_1 is zero, so the set M_2 of multiple points of f is not only nonempty, but gives a nontrivial element in the homology of M (dual to $w_{n-m}(\nu)$).

EXAMPLE (see [10]). Let f be a typical immersion of a compact surface M into $\mathbb{R}^3$. Then the number of triple points of this map is congruent mod 2 to the Euler characteristic of M. Indeed, formula (2) implies that this

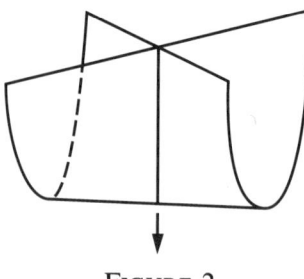

FIGURE 2

number is congruent mod 2 to the value of the square of the first Stiefel-Whitney class of the normal bundle of M at the fundamental cycle of M, and this value is always equal to the Euler characteristic.

Now let $f: M^2 \to \mathbb{R}^3$ be not necessarily an immersion, i.e., f may have several singular points. If f is in general position, all these singular points are Whitney umbrellas (see Figure 2). Let us displace the tip of each umbrella slightly from the surface $f(M)$ in the direction of the line of self-intersection. We obtain a set of points W in $\mathbb{R}^3 \setminus f(M)$. The number of these points is even: the line of self-intersection of $f(M)$ leaving any umbrella may terminate only at another umbrella. Consequently, $W = 0$ in $H_0(\mathbb{R}^3, \mathbb{Z}_2)$.

THEOREM (see [19]). *The number of triple points of a general map of a compact surface M into $\mathbb{R}^3$ is congruent mod 2 to the sum of the Euler characteristics of M and the linking number of the cycles $f(M)$ and W in $\mathbb{R}^3$.*

Complexes of multisingularities. We have just seen that the number of Whitney umbrellas of a map of a compact surface in N^3 is even. Similarly, the number of swallow-tails on a compact two-dimensional wave front is even (see Figure 3 of the lecture on hyperbolic operators in this volume). These remarks can be generalized to higher dimensions using the complex of multisingularities. This complex yields new bordism invariants and new restrictions on the coexistence of singularities.

Let Γ be a classification of maps $\mathbb{R}^m \to \mathbb{R}^n$, $m \le n$. Then points $y \in N_k$ of multiplicity k of any map $f: M^m \to N^n$ are also naturally classified in accordance with the Γ-classes of singularities of f at the points of the set $f^{-1}(y)$; to each class of this classification Γ^k on N_k corresponds an unordered set of k classes of the original classification Γ. (In particular, Γ^0 consists of one class, corresponding to the set $N = f(M)$.) Such sets are called *classes of multisingularities*. If Γ satisfies certain regularity conditions, then for almost any map $f: M \to N$ the partition of the manifold $N =$

$N_0 \cup N_1 \cup \cdots$ into the points of all possible classes Γ^k, $k = 0, 1$, is a regular partition, and one can construct chains and cycles in N from its elements. The homology of N that they define corresponds to the cohomology of the formal complex of multisingularities $\mathcal{N}(\Gamma)$.

By definition, this complex is the free $\mathbb{Z}_2$-module, generated by the multiclasses $\{\sigma_1, \ldots, \sigma_k\}$ of the classification Γ^k, $k = 0, 1, \ldots$. The dimension of a generator is the sum of the codimensions in N of the subsets $f(\sigma_i)$, $i = 1, \ldots, k$, for a typical map f having such singularities. The coboundary operators are defined by the mutual incidence geometry of the multiclasses: the incidence coefficient of multiclasses σ^1, σ^2 of codimensions $l, l+1$ is congruent mod 2 to the number of components of the set $\{\sigma^1\} \subset N$ in the neighborhood of any point of the set σ^2 of a typical map $f : M \to N$.

EXAMPLE. In the case of general maps $M^2 \to N^3$ the unique generator of the group $\mathcal{N}^2$ corresponds to the lines of self-intersection; generators of $\mathcal{N}^3$ are a Whitney umbrella and a triple point. Then δ (line of self-intersection)=(Whitney umbrella). Indeed, the set of double points in the neighborhood of a triple point is represented by an even number of branches (six), so that the incidence coefficient [line of self-intersection, triple point] is equal to zero. For Legendre maps $M^2 \to N^3$ the generators of the group $\mathcal{N}^2$ are a line of self-intersection and a cuspidal edge, the generators of the group $\mathcal{N}^3$ are a triple point, a transversal intersection of a cuspidal edge with a nonsingular part of the front and a swallow-tail. Then δ (cuspidal edge)=0, δ (line of self-intersection) = (swallow-tail).

MAIN THEOREM. 1. *Let the sum l of multisingularities $\sigma^1 + \cdots + \sigma^l$, $\sigma = \{\sigma_1^i, \ldots, \sigma_{k(i)}^i\}$, be a q-dimensional cocycle of the complex of multisingularities. Then for any nondegenerate map of a compact manifold M into N^n the union of the corresponding multisingularities in N^n forms an $(n-q)$-dimensional cycle mod 2. The class of this cycle is invariant to homotopies of the map.*

2. *If $n = q$, that is, the cycle is a set of points, then their sum mod 2 is a bordism invariant of maps in N^n.*

3. *If, in addition, $\sigma^1 + \cdots + \sigma^l$ is a coboundary of the complex $\mathcal{N}(\Gamma)$, then their number is always even.*

This theorem holds for any theory of singularities and smooth bordisms (Lagrange, Legendre, ordinary smooth maps...), the only restriction being $\dim M \leq \dim N$. Computations carried out in [22] for the complex of Legendre multisingularities (multisingularities of wave fronts) give the following new restrictions on the coexistence of singularities.

On a compact n-dimensional wave front in general position, the numbers of points of type A_3 for $n = 2$, of type A_5 for $n = 4$, and of type A_7 for

$n = 6$ are even. For $n = 5$ the number of points of type E_6 and the number of points where the front is pierced by its own one-dimensional stratum A_5 are equal modulo 2. The first three statements are easily carried over to the case of Lagrange singularities (see [22]).

As in the case of ordinary complexes of singularities, the actual calculations in complexes of multisingularities of general (non-Legendre) maps constitute a problem. In addition, one can define similar complexes of (co)oriented multisingularities.

References

1. V. I. Arnol'd, *On the characteristic class in quantization conditions*, Funktsional. Anal. i Prilozhen. **1** (1967), no. 1, 1–14; English transl. in Functional Anal. Appl. **1** (1967).
2. _____, *Lagrange and Legendre cobordisms*, Funktsional. Anal. i Prilozhen. **14** (1980), no. 3, 1–14; no. 4, 4–10; English transl. in Functional Anal. Appl. **14** (1980).
3. _____, *Singularities of systems of rays*, Uspekhi Mat. Nauk **38** (1983), no. 2, 77–147; English transl. in Russian Math. Surveys **39** (1983).
4. V. I. Arnol'd, A. N. Varchenko, and S. M. Guseĭn-Zade, *Singularities of differentiable maps*. I, "Nauka", Moscow, 1982; English transl., Birkhäuser, Basel, 1985.
5. V. I. Arnol'd, V. A. Vasil'ev, V. V. Goryunov, and O. V. Lyashko, *Singularities*. I, *Local and global theory*, Itogi nauki i tekhniki. Sovremennye Problemy Matematiki, Fundamental'nye Napravleniya, VINITI, Moscow; English transl., Encyclopedia Math. Sci., vol. 6, Springer-Verlag, Berlin, Heidelberg, and New York, 1988 (to appear).
6. M. Audim, *Cobordismes d'immersions lagrangiennes et legendriennes*, Travaux en Cours, Hermann, Paris, 1987.
7. J. M. Boardman, *Singularities of differentiable maps*, Inst. Hautes Études Sci. Publ. Math. **33** (1967), 21–57.
8. A. Borel and A. Haefliger, *La classe d'homologie fondamentale d'une espace analytique*, Bull. Soc. Math. France **89** (1961), 461–513.
9. Ya. M. Eliashberg, *Cobordisme des solutions de relations différentielles*, Seminaire Sud-Rhodanien de Geometrie, Travaux en Cours, vol. 1., Hermann, Paris, 1984, pp. 17–31.
10. T. Banchoff, *Triple points and singularities of smoothly immersed surfaces*, Proc. Amer. Math. Soc. **46** (1974), 402–406.
11. A. Haefliger and A. Kosinski, *Un théorème de Thom sur les singularités des applications*, Sem. H. Cartan in E.N.S., 1956/57, exposé 8.
12. V. P. Maslov, *Perturbation theory and asymptotical methods*, Izdat. Moskov. Gos. Univ., Moscow, 1965; French transl., Dunod, Paris, 1972.
13. J. Mather, *On Thom-Boardman singularities*, Dynamical Systems (J. Peixoto, ed.), Academic Press, New York, 1973, pp. 232–248.
14. J. Milnor and J. D. Stasheff, *Characteristic classes*, Ann. Math. Studies, vol. 76, Princeton Univ. Press, Princeton, N. J.; Univ. of Tokyo Press, Tokyo, 1974.
15. F. Pham, *Introduction à l'étude topologique des singularités de Landau*, Mémorial des Sciences Mathématiques, Fasc. 164, Gauthier-Villars, Paris, 1967.
16. I. R. Piorteous, *Simple singularities of maps*, Proc. of Liverpool Singularities Symposium. I, Lecture Notes in Math., vol. 192, Springer-Verlag, Berlin and New York, 1971, pp. 286–312.
17. F. Ronga, *Le calcul des classes duales aux singularités de Boardman d'ordre deux*, Comment. Math. Helv. **47** (1972), no. 1, 15–35.
18. _____, *On multiple points of smooth immersions*, Comment. Math. Helv. **55** (1980), no. 4, 521–527.
19. A. Szucs, *Surfaces in* $\mathbb{R}^3$, Bull. London Math. Soc. **18** (1986), no. 1, 60–66.

20. R. Thom, *Les singularités des applications différentiables*, Ann. Inst. Fourier (Grenoble) **6** (1956), 43–87.
21. V. A. Vasil'ev, *Characteristic classes of Lagrange and Legendre manifolds*, Funktsional. Anal. i Prilozhen. **15** (1981), no. 3, 10–22; English transl. in Functional Anal. Appl. **15** (1981), no. 3.
22. _____, *Lagrange and Legendre characteristic classes*, Adv. Studies Contemp. Math., vol. 3, Gordon and Breach, New York, 1988.

Translated by LEONID BRAILOVSKY

Lacunas of Hyperbolic Partial Differential Operators and Singularity Theory

V. A. VASSILIEV

Hyperbolic equations form a wide class of linear differential equations. The most famous representative of this class is the wave equation $\frac{\partial^2 u}{\partial t^2} - k \sum \frac{\partial^2 u}{\partial z_i^2} = 0$, which describes the propagation of waves with velocity k; by analogy, solutions of arbitrary hyperbolic equations are also called waves. An elementary wave, originating from an instantaneous point perturbation, has a singularity on a certain cone in space-time — the so-called wave front; it is analytic outside the cone and vanishes outside its convex hull. For example, the wave front of the wave equation is defined by the conditions $k^2 t^2 = \sum z_i^2$, $t \geq 0$. The topic of this lecture is the qualitative behavior of a wave in the vicinity of its front.

Even in the relatively simple case of the wave equation there are clearly various possible types of qualitative behavior. Thus, in our four-dimensional space-time (or any other 2ℓ-dimensional space, $\ell \geq 2$) a signal is noticed only at one instant, when it goes by the observer. On the other hand, in the odd-dimensional case the signal continues to be heard constantly after the time of encounter t_0 (with intensity proportional to $1/\sqrt{t^2 - t_0^2}$). The first situation is what enables us to communicate by means of sound; the second is responsible for the fact that the "acoustical layer" in the ocean, while an excellent transmitter of instantaneous signals, is unsuitable for the transmission of any compound information. Both variants of the behavior of sound waves have analogies in the case of arbitrary hyperbolic equations; in the language of the general theory one says that in the first case the inner component of the complement of the front is a lacuna, and in the second case, diffusion of waves occurs near the component; the outer component is a lacuna in any number of dimensions (and for any hyperbolic equation).

The condition that an elementary wave displays regular behavior ("sharpness") near the wave front is described in terms of singularities of smooth functions: it is equivalent to the condition that a certain vanishing cocycle in the cohomology of the nonsingular fiber of a singularity of the front-generating function should be trivial.

§1. Basic notions

1.1. Hyperbolic operators. Consider the linear space $\mathbb{R}^n_x$ with coordinates x_i, $i = 1, \ldots, n$, and let P be a differential operator with constant coefficients in $\mathbb{R}^n_x$, i.e., a finite sum of the type

$$\sum P_\alpha (-i\partial/\partial x_1)^{\alpha_1} \cdots (-i\partial/\partial x_n)^{\alpha_n},$$

where P_α are constants enumerated by multi-indices $\alpha = (\alpha_1, \ldots, \alpha_n)$. To any such operator we associate its characteristic polynomial $P = \sum P_\alpha \xi^\alpha$, $\xi^\alpha = \xi_1^{\alpha_1}, \ldots, \xi_n^{\alpha_n}$, in the coordinates ξ_i of some n-dimensional space $\mathbb{R}^n_\xi$. It is convenient to regard $\mathbb{R}^n_\xi$ as the dual of $\mathbb{R}^n_x$ with the duality defined by the formula $\langle (\xi_1, \ldots, \xi_n), (x_1, \ldots x_n) \rangle = \sum \xi_i x_i$. The order of the operator P is denoted by $\deg P$.

Let us consider the Cauchy problem for P in the half-space $x_1 \geq 0$. The choice of half-space singles out the coordinate x_1 (in the case of the wave equation x_1 is time); accordingly, the unit vector $\vartheta = (1, 0, \ldots, 0)$ is singled out in dual space $\mathbb{R}^n_\xi$.

Let $\overline{P}$ be the principal homogeneous part (highest-order terms) of the polynomial P. The equation $\overline{P}(\xi_1, \ldots \xi_n) = 0$ defines a conic hypersurface $A = A(P)$ in $\mathbb{C}^n_\xi$; the real part of this hypersurface is called the characteristic cone of P, denoted by $\operatorname{Re} A$.

DEFINITION. The operator P is said to be *hyperbolic in Petrovsky's sense* (or *strictly hyperbolic*) if the set $\operatorname{Re} A$ is nonsingular outside the origin and any straight line in $\mathbb{R}^n_\xi$ parallel to the unit vector ϑ which misses the origin cuts $\operatorname{Re} A$ in exactly $\deg P$ different points.

EXAMPLES. 1. The wave operator.

2. Any operator in $\mathbb{R}^2$ such that $P(\vartheta) \neq 0$.

3. A nondegenerate cubic curve in $\mathbb{R}P^2$ may have one or two components (and is represented either by both curves in Figure 1a or by the right-most curve only). The corresponding third-degree polynomial in $\mathbb{R}^3_\xi$ is never hyperbolic in the second case, while in the first case it is hyperbolic if and only if the unit vector ϑ is directed towards the interior of the conically shaped component of the surface $\operatorname{Re} A$.

Numerous other examples of hyperbolic operators are given in [15, 4].

The hyperbolic operators in Petrovsky's sense form an open domain in the space of operators of given order defined in $\mathbb{R}^n_x$.

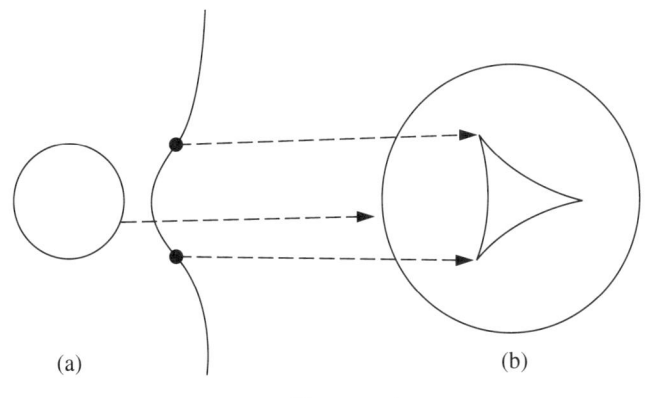

FIGURE 1

THEOREM (see [15]). *For any natural numbers k, n, the set of polynomials of degree k in $\mathbb{R}^n_\xi$ which are hyperbolic in Petrovsky's sense, relative to a fixed coordinate system, is the union of two connected components, each being contractible.*

More delicate properties of the domain of hyperbolicity are studied in [7].

1.2. The wave front. The basic geometrical object involved in the qualitative description of solutions of a hyperbolic equation is its wave front, which we shall now define. Let ξ be an arbitrary point of the cone $\operatorname{Re} A$, $\xi \neq 0$. We shall regard the tangent plane to $\operatorname{Re} A$ at this point as a subspace of $\mathbb{R}^n_\xi$. Consider the set of all vectors x in the initial space $\mathbb{R}^n_x$ orthogonal to the tangent plane, whose first coordinate x_1 is positive. The set of all such x for a given point ξ is a ray in $\mathbb{R}^n_x$.

DEFINITION. The union of all such rays for all points $\xi \in \operatorname{Re} A \setminus \{0\}$ is called *the wave front of the operator*, denoted by $W(P)$.

The wave front may have singularities even far from the origin: they correspond to flat points of $\operatorname{Re} A$, i.e., points at which the curvature form of $\operatorname{Re} A$ has rank less than $n-2$. For example, the projectivization of the wave front corresponding to the cubic surface in Figure 1a is shown in Figure 1b. Under this correspondence the points of inflection of the original curve go into cusps of the projectivization (the third point of inflection in Figure 1a is at infinity).

1.3. THEOREM (see [15, 4]). *If the operator P is hyperbolic in Petrovsky's sense, then*

(A) *P has a fundamental solution $E(P)$ whose support is a cone with vertex at 0, situated in the subspace $x_1 \geq 0$ and intersecting the plane $x_1 = 0$ only at the origin.*

(B) *This fundamental solution $E(P)$ is unique.*

(C) *$E(P)$ is analytic outside the surface $W(P)$ and vanishes outside the convex hull of $W(P)$.*

REMARK. (A) of Theorem 1.3 may be formalized as a definition of (nonstrict) hyperbolicity: An operator P is said to be hyperbolic if it has a fundamental solution whose support is contained in a cone with the properties described in (A). The wave front can also be defined for general hyperbolic operators (see [4]), and all the assertions of Theorem 1.3 remain valid. These operators have an explicit algebraic characterization as well (see [4]), but hyperbolicity is no longer dependent exclusively on the principal part of the operator: the lower-order terms must also be known. The hyperbolic, but not strictly hyperbolic polynomials form a set of positive codimension in the space of all polynomials; it is a subset of the closure of the set of strictly hyperbolic polynomials.

1.4. Sharpness, diffusion, lacunas.
Let P be a hyperbolic operator and W its wave front.

DEFINITION. (A) The fundamental solution $E(P)$ is said to be *holomorphically sharp* at a point y of W from the local (near y) component[1] ℓ of the complement of W if it has a holomorphic extension from ℓ to a neighborhood of y. Similarly, $E(P)$ is said to be C^∞-*sharp* if it has a C^∞-extension from ℓ to its closure $\bar{\ell}$. In these cases the component ℓ is called a *local (holomorhic or C^∞-) lacuna* of the operator P near y.

(B) If $E(P)$ is not sharp from ℓ we shall speak of *diffusion of waves* in ℓ.

(C) A component L of the complement of the front is called a *holomorphic lacuna* (C^∞-*lacuna*) if $E(P)$ is holomorphically (C^∞-) sharp from it at any point of its closure (or, equivalently, at the single point 0).

(D) If $E(P) = 0$ in L, then L is called a *strong lacuna* (see [4]) or simply a *lacuna* (see [15]).

EXAMPLE. The wave operator $P_n = (\partial^2/\partial t^2) - k^2 \sum_{i=2}^n \partial^2/\partial z_i^2$. It is well known that the corresponding fundamental solutions $E(P_n)$, $n = 2, 3, 4$, are given by the conditions

$$E(P_2) = \theta(kt - |z|)/(2k),$$
$$E(P_3) = \theta(kt - |z|)/\left(2\pi k \sqrt{k^2 t^2 - |z|^2}\right),$$
$$E(P_4) = \theta(t)\delta(k^2 t^2 - |z|^2)/(2\pi k),$$

where θ is the Heaviside function. Obviously, the inner component of the complement of the front is a strong lacuna when $n = 4$, a holomorphic but not a strong lacuna when $n = 2$, and diffusion occurs when $n = 3$. It turns out that when n is increased the qualitative situation for even n is the same as for $n = 4$, and for odd n as for $n = 3$. The exceptional situation when $n = 2$ is due to the fact that this value of n does not exceed the order of the operator.

---

[1] By "component" we shall always mean "arcwise connected component".

For the third-order equation in $\mathbb{R}^n_x$ the projectivization of whose wave front is described in Figure 1b, the innermost component (into which the inner "edges" protrude) is diffusive near any point of its boundary.

§2. Petrovsky's condition

Petrovsky, in [15], linked the property of a component L to be a lacuna with a topological condition, now known as Petrovsky's condition. This condition demands that a certain homology class—the Petrovsky class, which we shall soon define—should be trivial. Throughout this section, P is a strictly hyperbolic operator in $\mathbb{R}^n_x$.

Let $x \in \mathbb{R}^n_x$ be an arbitrary point of the component of the complement of the front W under consideration and $X \subset \mathbb{C}^n_\xi$ a hyperplane orthogonal to x. Let $\mathbb{C}P^{n-1}_\xi$ be the set of all one-dimensional complex subspaces of $\mathbb{C}^n_\xi$. Let X^* and A^* denote the hypersurfaces in $\mathbb{C}P^{n-1}_\xi$ obtained by projectivization of the hyperplane X and the cone A. The Petrovsky cycle we build is an $(n-2)$-dimensional cycle in the set $X^* - A^*$, which we are now going to describe.

Since x is not a point of the front, the hypersurfaces X^* and A^* intersect transversally near $\mathbb{R}P^{n-1}$ and their intersection is smooth. The set $X^* \cap \operatorname{Re} A^*$ of real points of the intersection is an $(n-3)$-dimensional submanifold $X^* \cap A^*$. If n is even it is an orientable manifold. A special choice of orientation, defined in [4], turns it into a cycle. The Petrovsky cycle $B(x) \subset X^* - A^*$ is defined as the image of this cycle under the tube map. (Let us recall the definition of the tube map. Consider a tubular neighborhood in X^* of the set of nonsingular points of the manifold $X^* \cap A^*$; stratify this neighborhood in some way into two-dimensional discs transversal to $X^* \cap A^*$. Then to any cycle $\triangledown$ contained in the smooth part of $X^* \cap A^*$ we can associate the union of the boundaries of the discs which meet $X^* \cap A^*$ in points of $\triangledown$; the complex orientation of X^*, $X^* \cap A^*$, and the original orientation of $\triangledown$ enable us to orient the tube.)

If n is odd, then far from A^* the cycle $B(x)$ is just the manifold $\operatorname{Re} X^*$ counted with multiplicity 2, but near A^* it splits into a pair of contours which encircle A^* and X^* on both sides: the nonreal part of the cycle is geometrically the same as the tube in $X^* - A^*$ around $X^* \cap \operatorname{Re} A^*$, as described above, but the two halves of the tube separated by the set $\operatorname{Re} X^*$ have opposite orientations. The cycle $B(x)$ for $n = 3$ is shown in Figure 2, with the set $A^* \cap X^*$ represented by crosses.

DEFINITION. The class $\beta(x)$ defined by $B(x)$ in $H_{n-2}(X^* - A^*)$—the homology group of $X^* - A^*$ with complex coefficients—is called the *Petrovsky class*. The condition $\beta(x) = 0$ is known as *Petrovsky's condition*.

It is easily seen that Petrovsky's condition is always fulfilled when $n = 2$; when $n = 3$ it is true if and only if $X^* \cap A^*$ has only real points (see Figure 2 on p. 30).

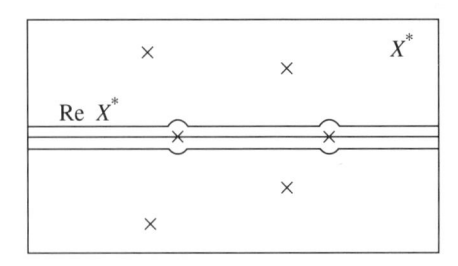

FIGURE 2

2.2. THEOREM (see [15, 4, 5, 10]). *If $\beta(x) = 0$, then the component L of the complement of the front that contains x is a holomorphic lacuna for P and for all operators with the same principal part $\overline{P}$ (if, in addition, $\deg P \leq n$ and $P = \overline{P}$, it is a strong lacuna). If A^* is a smooth set the converse is also true: if L is a C^∞-lacuna, then $\beta(x) = 0$ for every $x \in L$.*

In the homogeneous case $P = \overline{P}$, the first assertion of the theorem follows from the Herglotz-Petrovsky-Leray formula: this formula computes all the partial derivatives $D^\nu E$ of the fundamental solution $E = E(P)$ with $|\nu| \geq \deg P - n$ at a point $x \in L$ as integrals of differential $(n-1)$-forms $\Lambda(x, \nu, P)$, regular in the domain $\mathbb{C}P_\xi^{n-1} - X^* - A^*$, over the cycle $t\beta(x)$, which is contained in the same domain and obtained from $\beta(x)$ by the tube operation

$$t: H_{n-2}(X^* - A^*) \to H_{n-1}(\mathbb{C}P_\xi^{n-1} - X^* - A^*).$$

Consequently, if $\beta(x) = 0$, then $E(P)$ restricted to L is a polynomial of degree at most $\deg P - n$. The second assertion follows from the fact that (a) if L is a lacuna, then E/L is a polynomial; (b) in our case the tube operation is injective (in particular, if $t\beta(x) = 0$ then also $\beta(x) = 0$); and (c) for sufficiently large N the classes of all differential forms $\Lambda(x, \nu, P)$ such that $|\nu| = N$ generate the entire homology group $H^{n-1}(\mathbb{C}P_\xi^{n-1} - X^* - A^*)$ (see [5]). The case of nonhomogeneous P may be reduced to the homogeneous case (see [4, section 4.5]).

The following refinement of Theorem 1.3(B), is proved by similar methods.

2.3. THEOREM (see [5]). *If P is a hyperbolic operator and A^* a smooth set, then any point of the wave front $W(P)$ is a singular point of the fundamental solution $E(P)$. For almost every hyperbolic operator P with given principal part $\overline{P}$, the support of $E(P)$ is the convex hull of the wave front. If $n = 3$ these assertions are true for all hyperbolic operators, without exception.*

(For example, the wave operator with even $n \geq 4$ is one of the exceptions to the "almost every" rule.)

§3. The local Petrovsky condition

3.1. The same component of the complement of the wave front may be a local lacuna near some points of its boundary and a carrier of diffusion near others. The question of whether a component is a lacuna is equivalent to the question of whether it is a local lacuna near the origin. In order to investigate sharpness near other points of the front, Atiyah, Bott, and Gårding [5] introduced a local analogue of Petrovsky's condition which we shall now describe.

Let $y \neq 0$ be a point of the front $W = W(P)$, ℓ a component of the complement of W in a neighborhood of y, and $Y \subset \mathbb{C}P_\xi^n$ the projectivization of the plane $Y \subset \mathbb{C}_\xi^n$ orthogonal to y. If the point $x \in \ell$ is sufficiently close to y, the projection $Y \to X$ defines a homomorphism

$$P_x : H_{n-2}(Y^* - A^*) \to H_{n-2}(X^* - A^*) . \tag{1}$$

DEFINITION. The condition

$$\beta(x) \in P_x \left(H_{n-2}(Y^* - A^*) \right)$$

is called the *local Petrovsky condition*.

3.2. THEOREM (see [5]). *If the points of a local (near y) component ℓ of the complement of the front satisfy the local Petrovsky condition, then the fundamental solution $E(P)$ is holomorphically sharp from ℓ at y.*

The converse of Theorem 3.2 is almost always true:

3.3. THEOREM (see [9]). (1) *If the surface A^* is tangent to the plane Y^* near the set $\mathbb{R}P_\xi^{n-1}$ at finitely many points only, and $E(P)$ is holomorphically sharp from ℓ at y then the local Petrovsky condition holds for all $x \in L$.*

(2) *For almost every hyperbolic operator, the assumptions of* (1) *of our theorem hold for any point $y \neq 0$ of the front.*

Thus, for almost every hyperbolic operator, sharpness and the local Petrovsky condition are equivalent near all points of the front.

3.4. **Counterexample.** In the most general case the converse of Theorem 3.2 is false. Indeed, let $P = \xi(\xi_1^2 - \xi_2^2 - \xi_3^2)$. The front $W(P)$ is the union of a spherical cone and the ray spanned by the unit vector $(1, 0, 0)$. Sharpness near the ray follows from Hartog's theorem on removable singularities, but it is easy to see that the local Petrovsky condition is not satisfied.

§4. Geometry of lacunas near specific singularities of fronts

4.1. The local Petrovsky condition, and consequently also sharpness near points of the front, are defined by the local geometry of the front. The geometry is conveniently described in terms of (projective) generating functions, which we shall now define. For simplicity, we consider only the case in which the point $y \in W$ under consideration corresponds to a single point a of the set $\operatorname{Re} A$, i.e., the plane Y orthogonal to y is tangent to $\operatorname{Re} A$ along a single line. Choose an affine coordinate system $\eta_0, \eta_1, \ldots, \eta_{n-2}$ in $\mathbb{R}P_\xi^{n-1}$ with a center at a, relative to which the equation of the plane Y is $\eta_0 = 0$. Then the surface A^* is defined near a, by a condition $\eta_0 = f(\eta_1, \ldots, \eta_{n-2})$. The function f thus defined is called the *projective generating function* of A^* at a; obviously, $f(0) = \operatorname{grad} f(0) = 0$.

Let y^* be a point in $\mathbb{R}P^{n-1}$ orthogonal to the plane Y^*.

The neighborhood of the point y^* is the base of an important deformation of the function f. Indeed, for any point $u \in \mathbb{C}P^{n-1}$ close to y^*, the plane $U \subset \mathbb{C}P_\xi^{n-1}$ orthogonal to u is defined by a condition of the type $\eta_0 = u_0 + u_1\eta_1 + \cdots + u_{n-2}\eta_{n-2}$, where u_i are coefficients which may be regarded as local coordinates in $\mathbb{C}P^{n-1}$ near y^*. To every such point u we assign the function

$$f_u \equiv f(\eta_1, \ldots, \eta_{n-2}) - u_0 - u_1\eta_1 - \cdots - u_{n-2}\eta_{n-2}. \tag{2}$$

It is easily seen that the plane U is not transversal to the surface A near a if and only if the function f_u has a critical point near a with critical value zero. In other words, near y^* the projective wave front $W^*(P)$ coincides with the discriminant (bifurcation diagram of zeros) of the deformation (2) (see [1, 2], and also lectures of Givental' and Varchenko in this volume).

We have thus arrived at the following situation.

4.2. We are given a function $f : (\mathbb{C}^{n-2}, 0) \to (\mathbb{C}, 0)$, with an isolated singularity of finite multiplicity at 0, and a deformation $\{f_u\}$, $u \in \mathbb{C}^k$, of f where both f and $\{f_u\}$ are real (f has real values on $\mathbb{R}^{n-2}$, and the same is true for every function f_u for real u). We are going to examine the components of the complement to the discriminant of the deformation $\{f_u\}$ in $\mathbb{R}^k$.

4.3. In §4.1 we imposed additional conditions on f and f_u, in comparison with §4.2: f was required to be the projective generating function of an operator and f_u of the form (2), in particular, $k = n - 1$. It turns out that these additional conditions are superfluous: in §5 we shall redefine local lacunas in such a way that, even in the situation of §4.2, the question of whether a component of the complement of the discriminant is a lacuna is always meaningful; if f and f_u are obtained by the construction of §4.1, this definition will coincide with the previous one.

4.4. Investigation for sharpness near smooth points of the front. At a general point of the front, the singularity of f is a Morse singularity, i.e., its second differential is a nondegenerate quadratic form; we denote its indices of inertia by $i_\pm(a)$. In that case the front near a is a smooth manifold, which divides the neighborhood of a into two parts. In one of these parts there is a point u such that the corresponding plane U^* is defined by a condition $\eta_0 \equiv \varepsilon$, $\varepsilon \geq 0$. Denote this part by ℓ^+ and the other by ℓ^-.

THEOREM (see [12, 5]). *Let a be a general point of A^*. Then if n and $i_\pm a$ are even, both components $\ell^\pm$ are local lacunas; if n is even and $i_\pm(a)$ odd, there are no lacunas near y; if n is odd, then if $i_\pm(a)$ is even only ℓ^- is a lacuna, and if it is odd only ℓ^+ is a lacuna.*

4.5. Investigation for sharpness near an edge of regression and a swallow-tail.
At flat points of the surface $\operatorname{Re} A^*$ (i.e., points where the curvature form is degenerate), the generating function does not have a degenerate singularity (see, e.g., [1, 2]). Near the simplest singularities of this classification, those of types A_2 and A_3, the wave front is diffeomorphic to the product of a linear space and a semicubic parabola (respectively, a swallow-tail, i.e., the surface shown in Figure 3; points of type A_2 are represented in the figure by double lines). The condition for sharpness from a component of the complement of the front depends, once again, on the parity of n and that of the index i_+ of the quadratic part of the generating function; multiplying the coordinate η_0, if necessary, by -1 we get the modification A_3^+ of the singularity A_3; see the table below.

THEOREM (see [11]). *Near points of type A_2, for odd n and even i_+, one has sharpness from the component 2; for other parities of n and i_+, there is no sharpness. Near singularities of type A_3 the only local lacunas are the following: domain 3 in the case of odd i_+ and any n, and domain 2 in the case of even i_+ and odd n.*

4.6. For more complicated singularities, an analogous description of all local lacunas becomes more cumbersome. I shall confine myself, therefore, to some results concerning the number of lacunas.

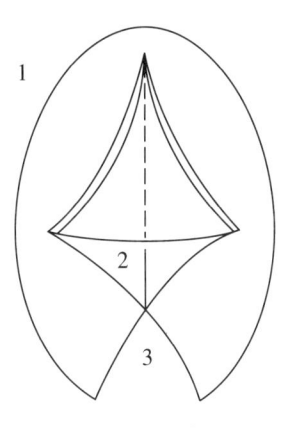

FIGURE 3

THEOREM (see [9, 3]). *In the neighborhood of any point where the wave front of the general hyperbolic operator has a singularity of one of the following types:*[2] A_k, D_k, E_k, P_8^1, X_9, X_{10}, J_{10}, J_{11}, $Y_{k,\mathbb{R}}$, Z_{11}, *the number of local lacunas is as specified in the table (or satisfies the inequality given there). For general fronts in* $\mathbb{R}^n$, $n \leq 7$, *only singularities of the types listed in the table may occur. (In the table, Q denotes a nondegenerate quadratic form in the additional variables* $z_{r+1}, \ldots, z_n$, *where r is the corank of the singularity, i.e., the deficiency of the curvature form of the surface* A^*, *listed in the third column of the table;* $i_\pm$ *is the positive index of inertia of the form Q. The column "$n \geq$" lists the minimal n such that the singularity in question will occur on a front in general position in* $\mathbb{R}^n$.)

TABLE 1

Type of lacunas	Normal form	corank	$n \geq$	n even, i_+ odd	n even, i_+ even	n odd, i_+ odd	n odd, i_+ even
A_1	Q	0	2	2	0	1	1
A_{2k}, $k \geq 1$	$x_1^{2k+1} + Q$	1	$2k+1$	0	0	1	0
$\pm A_{2k+1}$, $k \geq 1$	$\pm(x_1^{2k+2} + Q)$	1	$2k$	0	1	1	1
D_4^-	$x_1^2 x_2 - x_2^3 + Q$	2	5	0	3	1	1
D_{2k}^+, $k \geq 2$	$x_1^2 x_2 + x_2^{2k-1} + Q$	2	$2k+1$	0	0	1	1
D_{2k}^-, $k \geq 3$	$x_1^2 x_2 - x_2^{2k-1} + Q$	2	$2k+1$	0	2	1	1
$\pm D_{2k+1}^+$, $k \geq 2$	$\pm(x_1^2 x_2 + x_2^{2k} + Q)$	2	$2k+2$	0	0	1	1
$\pm E_6$	$\pm(x_1^3 + x_2^4 + Q)$	2	7	0	0	1	1
E_7	$x_1^3 + x_1 x_2^3 + Q$	2	8	0	0	1	1
E_8	$x_1^3 + x_2^5 + Q$	2	9	0	0	1	1
P_8^1	$x_1^3 + x_2^3 + x_3^3 + ax_1 x_2 x_3 + Q$, $a > -3$	3	8	0	0	≥ 1	0
$\pm X_9$	$\pm(x_1^4 + ax_1^2 x_2^2 + x_2^4 + Q)$, $a > -2$	2	9	≥ 1	0	≥ 2	0
X_9^1	$x_1 x_2(x_1^2 + ax_1 x_2 + x_2^2) + Q$, $|a| < 2$	2	9	0	0	0	0
X_9^2	$x_1 x_2(x_1 + x_2)(x_1 + ax_2) + Q$, $a \in (0,1)$	2	9	0	≥ 2	0	0
J_{10}^3	$x_1(x_1 - x_2^2)(x_1 - ax_2^3) + Q$, $a \in (0,1)$	2	10	≥ 0	≥ 1	0	0
J_{10}^1	$x_1(x_1^2 + ax_1 x_2^2 + x_2^4) + Q$, $|a| < 2$	2	10	≥ 0	≥ 0	0	0
$\pm J_{11}$	$\pm(x_1^3 + x_1^2 x_2^2 + ax_1^7 + Q)$, $a > 0$	2	11	≥ 0	≥ 0	≥ 0	0
$\pm P_{k,\mathbb{R}}$, $k \geq 5$	$\pm[(x_1^2 + x_2^2)^2 + ax_1^k + Q]$, $a > -3$	2	$k+5$	≥ 1	0	≥ 2	$0, k \geq 7$; $\geq 0, k = 5$
Z_{11}	$x_1^3 x_2 + x_2^5 + ax_1 x_3^4 + Q$	2	11	0	≥ 0	≥ 0	≥ 0

[2] The notation is taken from the classification tables in [1, 2].

CONJECTURE. *All entries in the table involving the symbol "≥ 0" should be replaced by zeros.*

The theorem is proved in [12] for A_1, in [11] for A_2 and A_3, and in [9] for A_k ($k \geq 4$), D_k, E_k. The assertions about the other singularities were established by computer (see [3]); the numerical work in question provides serious grounds in favor of our conjecture.

For a continuation of the table, see [3].

All the results of §4 (and Theorem 3.3) were established by means of local singularity theory (see [1–3]); in the following section we present the "principal actor" in the proofs of these results.

§5. The local Petrovsky cycle

Let f, $\{f_u\}$ be as in §4.2. Let $u \in \mathbb{R}^k$ be a nondiscriminant value of the parameter, i.e., the set $f_u^{-1}(0)$ is nonsingular in some prescribed spherical neighborhood B of the origin. According to [1, 2], the manifold $f_u^{-1}(0) \cap B$ (which we denote by V_u) has the following medium dimension homology groups: $\tilde{H}_{n-3}(V_u) \cong H_{n-3}(V_u, \partial V_u) \cong \mathbb{Z}^\mu$, where μ is the Milnor number of the singularity of f, and the "tilde" on H means that we are considering reduced homology modulo a point in the case of absolute homology, modulo the fundamental cycle in the relative case. Two cycles, the even and odd local Petrovsky cycles, are defined in the group $\tilde{H}_{n-3}(V_u, \partial V_u)$. The even Petrovsky cycle is defined by the set of real points of V_u, oriented in such a way that at any of its points the frame (grad f_u, the positively oriented tangent frame of the cycle $V_u \cap \mathbb{R}^{n-2}$) defines the positive orientation of the space $\mathbb{R}^{n-2}$. To define the odd cycle, Petrovsky pointed out that Levi's tube operation (see §2) defines an isomorphism $t : \tilde{H}_{n-3}(V_u, \partial V_u) \to H_{n-2}(B - V_u, \partial B - V_u)$ (see [9, Lemma 2.4.2]). In the above-mentioned group we define a relative cycle analogous to the odd Petrovsky cycle of §2: far from V_u it coincides with the set $\mathbb{R}^{n-2} \cap B$, counted with double multiplicity (endowed with the orientation of $\mathbb{R}^{n-2}$); near V_u it encircles it from both sides in the complex domain (see Figure 2). The odd Petrovsky cycle is defined as the preimage of this cycle under the tube isomorphism t.

DEFINITION. If n is even (odd), a component of the complement of the discriminant of a singularity of f is called a *homological local lacuna* if, for any point u of the component, the even (odd) local Petrovsky cycle in $H_{n-3}(V_u, \partial V_u)$ is trivial.

THEOREM (see [9]). *Let f be the projective generating function of a hyperbolic operator, having a singularity of finite multiplicity at $0 \in \mathbb{C}^{n-1}$, and let $\{f_u\}$ be the deformation of f as in* (2). *Then for any component of the complement of the discriminant of f_u (= the wave front) the following conditions*

are equivalent:
  (a) *the local Petrovsky condition for the points of the component*;
  (b) *the fundamental solution is sharp from the component*;
  (c) *the component is a homological local lacuna*.

For further information the reader is referred to [9, 3], where, among other things one can find:

— expressions for both local Petrovsky classes for real nondiscriminant Morsifications of any real singularity of finite multiplicity, in terms of standard topological characteristics—the form of the intersection in the homology groups of the manifold V_u and the Morse indices of the real singularities of f_u;

— the behavior of local Petrovsky classes upon passage from f to stably equivalent functions, i.e., functions of the form $f(\eta_1, \dots, \eta_{n-2}) \pm \xi_i^2$, $\pm \xi^2$;

— the behavior of these classes under all standard surgery modifications of the Morsification of f_u.

In conclusion, we mention that it is also possible to define a local Petrovsky class near points of the front corresponding to nonisolated singularities: it is a cohomology class in a certain bundle complex with support in the nontransversality set of A^* and Y^*.

§6. Equations with variable coefficients

All the results of §§ 3–5 have natural generalizations to hyperbolic operators with variable coefficients: $P = \sum P_\alpha(x)(\partial/\partial x)^\alpha$, $P_\alpha \in C^\infty(\mathbb{R}_x^n)$. In particular, one can also define sharpness and formulate the local homological Petrovsky condition. Without formalizing the concepts, we present only the main result of the theory.

THEOREM (see [8,11]). *Near the points of the wave front of a strictly hyperbolic operator corresponding to isolated singularities of the generating function, the local Petrovsky condition implies sharpness of the fundamental solution.*

This theorem was proved in [11] in the case of the analytic coefficient, and in the neighborhood of singularities of type A_k for the C^∞ case. There the author also outlined an approach to the proof of the theorem as just formulated, which was subsequently implemented in [8].

REFERENCES

1. V. I. Arnol'd, A. N. Varchenko, and S. M. Guseĭn-Zade, *Singularities of differentiable maps*, vol. II, "Nauka", Moscow, 1984; English transl., Birkhäuser, Basel, 1988.
2. V. I. Arnol'd, V. A. Vasil'ev, V. V. Goryunov, and O. V. Lyashko, *Singularities*. I, *Local and global theory*, Itogi Nauki i Tehniki. Sovremennye Problemy Matematiki. Fundamental'nye Napravleniya, vol. 6, VINITI, Moscow, 1988; English transl., Encyclopedia Math. Sci., vol. 6, Springer-Verlag, Berlin, Heidelberg, and New York (to appear).
3. _____, *Singularities*. II, *Classification and applications*, Itogi Nauki i Tehniki. Sovremennye Problemy Matematiki. Fundamental'nye Napravleniya, vol. 39, VINITI, Moscow, 1989; English transl., Encyclopedia Math. Sci., vol. 39, Springer-Verlag, Berlin, Heidelberg, and New York (to appear).

4. M. F. Atiyah, R. Bott, and L. Gårding, *Lacunas for hyperbolic differential operators with constant coefficients.* I, Acta Math. **124** (1970), 109–189.
5. _____, *Lacunas for hyperbolic differential operators with constant coefficients.* II, Acta Math. **131** (1973), 145–206.
6. V. A. Borovikov, *Fundamental solutions of linear partial differential equations with constant coefficients*, Trudy Moskov. Mat. Obshch. **8** (1959), 199–257. (Russian)
7. A. D. Vainshtein and B. Z. Shapiro, *Singularities of the boundary of the domain of hyperbolicity*, Itogi Nauki i Tekhniki. Sovremennye Problemy Matematiki. Noveĭshie Dostizheniya, vol. 33, VINITI, Moscow, 1988, pp. 193–216; English transl. in J. Soviet Math. **52** (1990).
8. A. N. Varchenko, *On normal forms of nonsmoothness of solutions of hyperbolic equations*, Izv. Akad. Nauk SSSR Ser. Mat. **51** (1987), no. 1, 114–131; English transl. in Math. USSR-Izv. **30** (1988).
9. V. A. Vasil′ev, *Sharpness and the local Petrovskiĭ condition for strictly hyperbolic operators with constant coefficients*, Izv. Akad. Nauk SSSR Ser. Mat. **50** (1986), no. 2, 242–283; English transl. in Math. USSR-Izv. **28** (1987).
10. A. M. Gabrielov, *A proof of I. G. Petrovsky's theorem*, Appendix to: I. G. Petrovskiĭ, Selected Works, vol. 1, "Nauka", Moscow, 1986. (Russian)
11. L. Gårding, *Sharp fronts of paired oscillatory integrals*, Publ. Res. Inst. Math. Sci. **12** (1977), 53–68.
12. A. M. Davydova, *Sufficient condition for the absence of lacunas*, Izd. Moskov. Gos. Univ., Moscow, 1945. (Russian)
13. J. Leray, *Un prolongement de la transformation de Laplace (Problème de Cauchy. IV)*, Bull. Soc. Math. France **90** (1962), 39–156.
14. W. Nuji, *A note on hyperbolic polynomials*, Math. Scand. **23** (1968), no. 1, 69–72.
15. I. G. Petrovskiĭ(= Petrovsky), *On the diffusion of waves and lacunas for hyperbolic equations*, Rec. Math. (= Mat. Sbornik) **17** (1945), 289–370.

Translated by TAMAR BURAK

Reflection Groups in Singularity Theory

A. B. GIVENTAL'

Introduction

These notes are concerned mainly with the numerous manifestations of the so-called A, D, E -classification. Dynkin diagrams can be used to classify crystallographic groups, simple Lie algebras, finite quaternion groups, singularities of wave fronts and caustics, simple categories of linear spaces—and probably many other things. Our attention will be concentrated on simple constructions that relate them to other simple objects. In §§1, 2, 3 we shall construct a commutative diagram

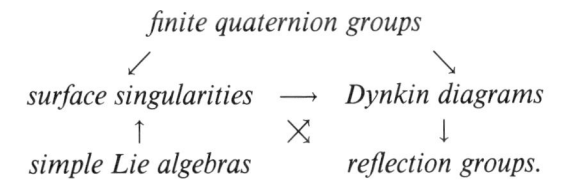

$$\begin{array}{ccc} & \textit{finite quaternion groups} & \\ \swarrow & & \searrow \\ \textit{surface singularities} & \longrightarrow & \textit{Dynkin diagrams} \\ \uparrow & \times & \downarrow \\ \textit{simple Lie algebras} & & \textit{reflection groups.} \end{array}$$

As one application of these constructions we shall obtain a description of singularities and their discriminants as orbit spaces of symmetry groups of suitable regular polyhedra. In §§4 and 5 we shall try to extend this result to nonsimple surface singularities, actually doing so for quasihomogeneous unimodal singularities, which come next in order of complexity. Here elliptic curves, Fuchsian groups, automorphic functions of several variables, complex crystallographic groups, and Torelli theorems for $K3$-surfaces come into the picture.

§1. The A, D, E - classification

1.1. Weyl groups [1]. Consider the lattice $\mathbb{Z}^\mu$ equipped with an integer-valued negative definite scalar product $\langle\,,\,\rangle$. The *Weyl group* is the finite group W of transformations of this lattice generated by reflections in hyperplanes orthogonal to the basis vectors $(\sigma_1, \ldots, \sigma_\mu)$ of the lattice.

Every Weyl group can be decomposed as a product of irreducible factors. The irreducible Weyl groups can be explicitly listed. They are represented by the following symbols (4 sequences + 5 exceptional groups):

$$A_\mu \ (\mu \geq 1), \ B_\mu \ (\mu \geq 2), \ C_\mu \ (\mu \geq 3), \ D_\mu \ (\mu \geq 4),$$
$$E_6, \ E_7, \ E_8, \ F_4, \ G_2.$$

EXAMPLE. A_μ is the permutation group of the coordinates $(x_0, \ldots, x_\mu)$ of the lattice $\mathbb{Z}^{\mu+1}$; it acts on the invariant hyperplane $\sum x_i = 0$.

The scalar product $\langle \, , \, \rangle$ on $\mathbb{Z}^\mu$ is defined by a *Dynkin diagram*. Diagrams A, D, E look like this:

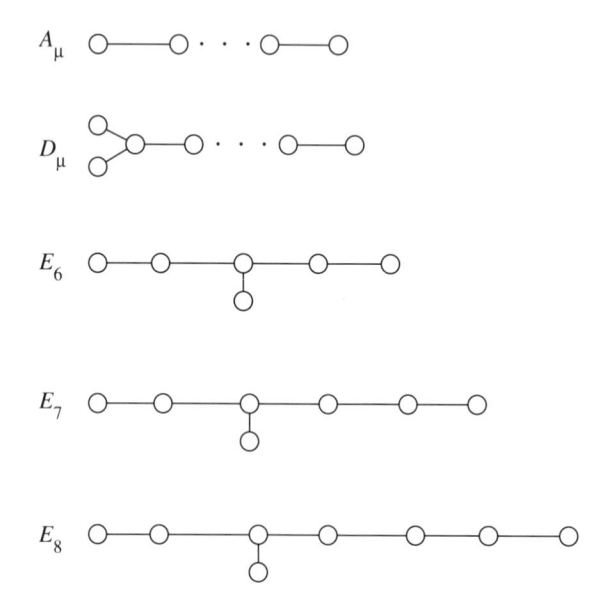

A_μ ○——○ · · · ○——○

D_μ (diagram)

E_6 (diagram)

E_7 (diagram)

E_8 (diagram)

The vertices of the diagram represent basis vectors; the scalar product of a vector with itself is $\langle \sigma, \sigma \rangle = -2$. The edges define angles between vectors: if there is no edge, the vectors are orthogonal: $\langle \sigma_i, \sigma_j \rangle = 0$; if there is an edge, the angle is $120°$ ($\langle \sigma_i, \sigma_j \rangle = 1$).

The Weyl groups are groups of symmetries of regular integral polyhedra. Thus, A_μ is the symmetry group of a μ-dimensional simplex, C_μ of a μ-dimensional cube, B_μ of its dual—a μ-dimensional "octahedron" (so that the groups are isomorphic and only the lattices are different). A polyhedron of type D can be described as follows: its vertices are the vertices $(\varepsilon_1, \ldots, \varepsilon_\mu)$ of the unit cube ($\varepsilon_i = \pm 1$) for which $\varepsilon_1 \cdots \varepsilon_\mu = 1$.

The symmetry group of a regular (not necessarily integral) polyhedron in Euclidean space $\mathbb{R}^\mu$ is called a *Coxeter group*.[1] Each Coxeter group decomposes into a product of irreducible factors. The list of irreducible Coxeter groups includes the previous list plus the symmetry groups H_2, H_3, H_4 of the regular pentagon in $\mathbb{R}^2$, the icosahedron in $\mathbb{R}^3$ and the 600-hedron

---

[1] On the occurrence of these groups in singularity theory and geometrical optics, see [17,18].

in $\mathbb{R}^4$, as well as the symmetry groups $I_2(p)$ of the regular p-gons, $p \geq 7$. In addition, in the list of Weyl groups one has the identification $B_\mu = C_\mu$.

1.2. Simple Lie algebras [2].
Simple finite-dimensional complex Lie algebras are also classified by Dynkin diagrams A, B, C, D, E, F, G. The classical simple Lie algebras are arranged in sequences:
$$A_\mu = sl_{\mu+1}, \quad B_\mu = so_{2\mu+1}, \quad C_\mu = sp_{2\mu}, \quad D_\mu = so_{2\mu}.$$
We recall the construction of the correspondence
$$\text{Lie algebra} \longmapsto \text{Weyl group}.$$
Let $\mathfrak{h}$ be a Cartan subalgebra of a simple Lie algebra $\mathfrak{g}$ (i.e., the Lie algebra of a maximal torus in the corresponding Lie group G). The normalizer of the maximal torus, factorized by the torus itself, acts in $\mathfrak{h}$ as a reflection group. This group is precisely the Weyl group. The scalar product $\langle \, , \, \rangle$ is defined as the restriction to $\mathfrak{h}$ of the Killing form:
$$\langle x, y \rangle = \text{tr}(\text{ad}_x \cdot \text{ad}_y).$$
The lattice $\mathbb{Z}_\mu \subset \mathfrak{h}$ is the dual of the lattice generated in $\mathfrak{h}^*$ by the characters of the adjoint representation
$$\text{ad}_\mathfrak{g} : \mathfrak{h} \to \text{End}\,\mathfrak{g}$$
of the Cartan subalgebra of $\mathfrak{g}$.

Conversely, the corresponding Lie algebra may be reconstructed from the Dynkin diagram in terms of generators and relations (see [2]).

EXAMPLE. A_μ. The Lie algebra $sl_{\mu+1}$ is the algebra of traceless matrices of order $\mu + 1$. The Cartan subalgebra is the set of diagonal matrices $\text{diag}(x_0, \ldots, x_\mu)$, $\sum x_i = 0$. Its normalizer in $SL_{\mu+1}$ acts by permutations of diagonal elements. The elementary matrices e_{ij} are the eigenvectors of the adjoint action of the torus. The corresponding characters are $x_i - x_j$. The scalar product of an element with itself, $\langle x, x \rangle = \sum (x_i - x_j)^2$, is proportional to $\sum x_i^2$.

1.3. Critical points of functions [3, 4, 20].
Let $f : (\mathbb{C}^3, 0) \to (\mathbb{C}, 0)$ be a germ of a holomorphic function at an isolated critical point: $d_0 f = 0$. Two such germs are said to be *equivalent* if one of them can be mapped onto the other by a germ of a diffeomorphism of the domain. The equivalence classes may form continuous families, that is, they may depend on invariants, called *moduli*. However, the beginning of the classification is discrete. Discrete equivalence classes are called *simple*. Here is a list of the normal forms of simple singularities for functions of three variables (x, y, z):
$$A_\mu \ (\mu \geq 1): x^{\mu+1} - yz, \quad D_\mu \ (\mu \geq 4): x^2 y - y^{\mu-1} + z^2,$$
$$E_6: x^3 + y^4 + z^2, \quad E_7: x^3 + xy^3 + z^2, \quad E_8: x^3 + y^5 + z^2.$$
The list for more variables is the same, except that nondegenerate quadratic forms of the additional variables must be added to the normal forms.

1.4. Finite quaternion groups [6, 19].

Let $\mathbb{H}$ be the quaternion algebra $\mathbb{H} \simeq \mathbb{C}^2 \simeq \mathbb{R}^4$:

$$q = z_1 + z_2 j = (a + bi) + (c + di)j = a + bi + cj + dk,$$
$$i^2 = j^2 = k^2 = -1, \quad ij = k, \quad jk = i, \quad ki = j.$$

The quaternions of unit length form a group Sp_1, which is isomorphic to SU_2 (in the complex representation

$$q \mapsto \begin{pmatrix} z_1 & -\bar{z}_2 \\ z_2 & \bar{z}_1 \end{pmatrix}, \quad |z_1|^2 + |z_2|^2 = 1).$$

Let Γ be a finite subgroup of Sp_1. The list of all these subgroups up to conjugation is

$$A_\mu \ (\mu \geq 1), \quad D_\mu \ (\mu \geq 4), \quad E_6, \quad E_7, \quad E_8,$$

where A_μ is a cyclic subgroup of order $\mu + 1$, D_μ, E_6, E_7, E_8 are the *binary* groups of a dihedron (regular $(\mu - 2)$-gon in space), a tetrahedron, an octahedron, and an icosahedron, respectively.

The binary group of a regular polyhedron in $\mathbb{R}^3$ is defined as follows. The quotient group $SU_2/(\pm 1)$ by the center is isomorphic to SO_3. The group of *rotations* of a polyhedron is a subgroup of SO_3. Its inverse image in SU_2 is just the binary group of the polyhedron.

EXAMPLES. (1) The group of *symmetries* of a dihedron with $\mu \geq 5$ is the Coxeter group $A_1 \oplus I_2(\mu - 2)$.

(2) If $\mu = 4$, the group of symmetries of the dihedron is $A_1 \oplus A_1 \oplus A_1$. The corresponding binary group consists of the quaternions ± 1, $\pm i$, $\pm j$, $\pm k$.

(3) The binary groups E_6, E_7, E_8 appear in Coxeter's classification under the names A_3, B_3, H_3 of the groups of symmetries of the corresponding polyhedra.

(4) *Description of the Coxeter group H_4.* The binary group of an icosahedron is a $\mathbb{Z}_2$-extension of its group of rotations—a simple group of 60 elements. The 120 elements of the binary group are 120 points on the unit sphere in the four-dimensional space of quaternions. Their convex hull is a regular polyhedron in $\mathbb{R}_4$, whose symmetry group is just the Coxeter group H_4. The subgroup SH_4 (of index 2) of rotations of this polyhedron can be described as follows. A pair (u_1, u_2) of unit quaternions acts by rotations in the quaternion space: $q \mapsto u_1^{-1} q u_2$. This action generates an exact sequence

$$1 \to (\pm 1) \to SU_2 \times SU_2 \to SO_4 \to 1.$$

The "restriction" of this exact sequence to the binary group Γ of the icosahedron,

$$1 \to (\pm 1) \to \Gamma \times \Gamma \to SH_4 \to 1,$$

gives the required description of SH_4 and shows that the order of the group H_4 is $|\Gamma|^2 = 14400$.

EXERCISE. Which polyhedra and Coxeter groups are obtained by an analogous construction for the other binary groups D_μ, E_6, E_7?

1.5. Quotient-singularities [6, 19].

The following construction defines a correspondence

$$\text{finite quaternion groups} \longleftrightarrow \text{simple critical points of functions.}$$

Let $\Gamma \subset SU_2$ be a finite subgroup. It acts in $\mathbb{C}^2$ together with SU_2.

THEOREM (Klein). *The surface $\mathbb{C}^2/\Gamma$ has an isolated simple singularity at zero of the same type as Γ.*

More precisely, the theorem states that the algebra of Γ-invariant polynomials on $\mathbb{C}^2$ has 3 generators x, y, z, satisfying a relation $f(x, y, z) = 0$, where f is a polynomial on $\mathbb{C}^3$ with an isolated simple critical point 0.

EXAMPLE. A_μ. Γ is generated in SU_2 by the matrix $\operatorname{diag}(\varepsilon, \varepsilon^{-1})$, where ε is a $(\mu+1)$th primitive root of 1. The basis invariants are expressed in terms of the variables (w_1, w_2) in $\mathbb{C}^2$ as follows: $x = w_1 w_2$, $y = w_1^{\mu+1}$, $z = w_2^{\mu+1}$. They are related by $x^{\mu+1} = yz$.

1.6. Extended Dynkin diagrams [21].

McKay's construction, to be described below, defines a map

$$\text{finite quaternion groups} \longrightarrow \text{Dynkin diagrams,}$$

justifying the notation A, D, E for binary groups and simple singularities.

Let Γ be a finite group and R a complex representation of Γ. We construct a graph as follows. The vertices of the graph are indexed by the irreducible representations $V_0 = 1, V_1, \ldots, V_\mu$ of Γ. Let

$$R \otimes V_i = \bigoplus_j m_{ij} V_j$$

be the decomposition of the tensor product into irreducible components. Then we connect vertex i to vertex j by m_{ij} directed edges. Two edges going in opposite directions are then replaced by an undirected edge.

THEOREM (McKay). *Let Γ be a finite quaternion group and R its standard representation in $\mathbb{C}^2$. Then the corresponding graph is the extended Dynkin diagram of the same type.*

The *extended Dynkin diagrams* of types A, D, E look like this:

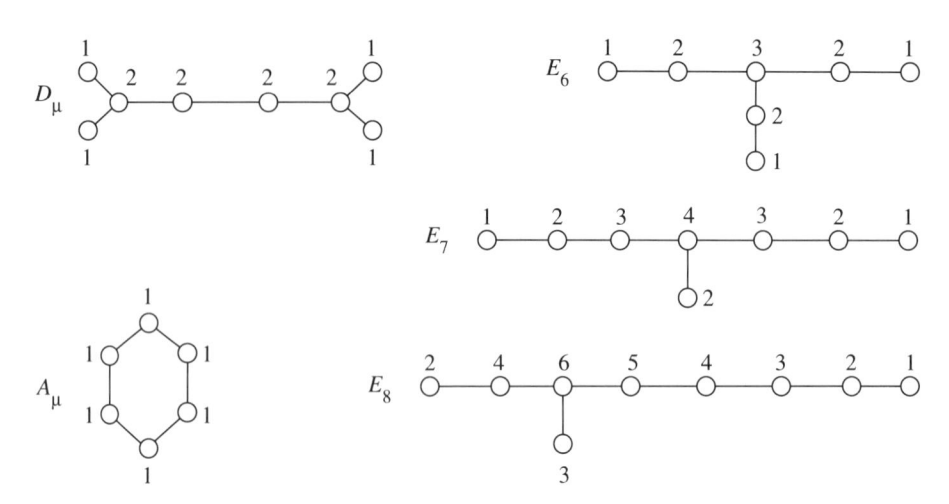

[Diagrams of extended Dynkin diagrams D_μ, A_μ, E_6, E_7, E_8 with vertex weights]

The number of vertices is $\mu + 1$. The distinguished "superfluous" vertex corresponds to the identity representation. The numbers indicate the dimensions of the irreducible representations. (Recall that a graph with numbers, weights, assigned to its vertices is called a *weighted* graph.)

EXAMPLE. A_μ. In this case the group Γ is commutative. Its irreducible representations V_k are one-dimensional and are determined by action of the generator $e \mapsto \varepsilon^k$. The standard representation R is reducible and isomorphic to $V_1 \oplus V_\mu$. Hence $R \otimes V_k = V_{k+1} \oplus V_{k-1}$ $(k \in \mathbb{Z}/(\mu+1)\mathbb{Z})$.

The proof of McKay's theorem is based on a purely combinatorical lemma. We call a weighted graph an *r-graph* if the sum of weights of the vertices incident to a given vertex (counting multiplicities of edges) is r times the weight of that vertex.

LEMMA. *The connected weighted undirected 2-graphs without loops are exactly the extended Dynkin diagrams A, D, E.*

EXERCISES. (1) Prove that the graph of the pair (Γ, R), weighted by the dimensions of irreducible representations, has the following properties:

  (i) it is an r-graph with $r = \dim R$;
  (ii) it is undirected if and only if $R \simeq R^*$ is a self-dual representation;
  (iii) it is connected if and only if R is a faithful representation of Γ.

(2) Deduce from this (and the Lemma) that (Γ, R) is an extended Dynkin diagram if and only if $\Gamma \subset SU_2$ and R is the standard representation.

(3) Let R be the standard representation of an infinite compact subgroup $\Gamma \subset SU_2$. What are the weighted graphs (Γ, R)? (Note that Γ is one of the three groups $T \subset D \subset SU_2$, where T is a circle, D the infinite dihedral group of the "regular ∞-gon", i.e., the circle. The corresponding diagrams are denoted by A_∞, D_∞, E_∞.)

§2. Discriminant

2.1. Chevalley's theorem [1]. Let W be a Coxeter group. It acts in the complexification $\mathbb{C}^\mu$ of the Euclidean space $\mathbb{R}^\mu$.

CHEVALLEY'S THEOREM. *The orbit space $\mathbb{C}^\mu/W$ is isomorphic to μ-dimensional complex space.*

In other words, the algebra of W-invariant polynomials in $\mathbb{C}^\mu$ is itself isomorphic to the algebra of polynomials in μ generators.

Regular orbits (that is, orbits in general position) consist of $|W|$ points. The nonregular orbits form a singular hypersurface $A \subset \mathbb{C}^\mu/W$, called the *discriminant of the Coxeter group*; it coincides with the image of the "mirrors" (i.e., the fixed hyperplanes of the reflections in W) under the projection $\mathbb{C}^\mu \to \mathbb{C}^\mu/W$.

EXAMPLES. (1) The action of A_μ in $\mathbb{C}^\mu$ coincides with that of the group of permutations of the roots of the polynomials

$$(x - x_0) \cdots (x - x_\mu), \qquad \sum x_i = 0,$$

in the root space. The symmetric functions $\sigma_2, \ldots, \sigma_{\mu+1}$ form a basis of the algebra of W-invariant polynomials. Thus the orbit space is isomorphic to the space of polynomials

$$x^{\mu+1} + \lambda_1 x^{\mu-1} + \cdots + \lambda_\mu.$$

The discriminant of A_μ is the set of all polynomials with multiple roots, i.e., the discriminant of the space of polynomials.

(2) The discriminants of A_3, B_3, H_3 look as follows [3, 17, 31]:

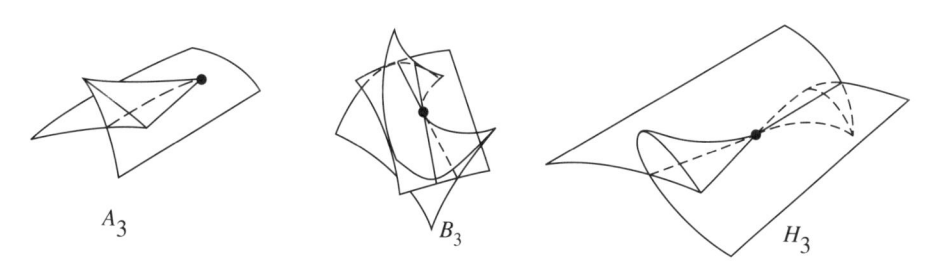

A_3 $\qquad B_3$ $\qquad H_3$

Cutting these surfaces by a plane in general position through the origin, we obtain a plane curve with a simple singularity:

$$x^3 = y^4, \qquad x^3 = xy^3, \qquad x^3 = y^5.$$

EXERCISE. Deduce from this Klein's theorem for the quaternion groups E_6, E_7, E_8.

2.2. **Invariants of the adjoint representation** [2]. Let $\mathfrak{g}$ be a semisimple complex Lie algebra of rank μ (recall that the rank is the dimension of a Cartan subalgebra) and $\mathrm{Ad}: G \to \mathrm{GL}(\mathfrak{g})$ its adjoint representation.

THEOREM. *The algebra of* Ad-*invariant polynomials is isomorphic to the algebra of polynomials in μ generators.*

These generators determine a projection $\mathfrak{g} \to \mathfrak{g}/G \simeq \mathbb{C}^\mu$ onto the "orbit space" of the adjoint representation. The critical values of the projection form a singular hypersurface in $\mathbb{C}^\mu$, called the *discriminant of* $\mathfrak{g}$.

EXAMPLE. A_μ. Under the projection $\mathfrak{g} \to \mathfrak{g}/G$, a matrix of order $\mu+1$ with zero trace is mapped to its characteristic polynomial. If all its roots are simple, then the typical fiber is one orbit of the adjoint representation (of all matrices with a given simple spectrum) and is nonsingular. Hence the discriminant of the algebra $sl_{\mu+1}$ is just the discriminant in the space of polynomials

$$x^{\mu+1} + \lambda_1 x^{\mu-1} + \cdots + \lambda_\mu.$$

THEOREM. *Restriction to the Cartan subalgebra defines an isomorphism of the algebra of* Ad-*invariant polynomials on a semisimple Lie algebra onto the algebra of invariants of its Weyl group.*

COROLLARY 1. *The following diagram is commutative*:

$$\begin{array}{ccc} \mathfrak{h} & \hookrightarrow & \mathfrak{g} \\ \downarrow & & \downarrow \\ \mathfrak{h}/W & = & \mathfrak{g}/G \simeq \mathbb{C}^\mu. \end{array}$$

COROLLARY 2. *The discriminant of a semisimple Lie algebra is isomorphic to the discriminant of its Weyl group.*

2.3. **Versal deformations** [3, 4]. Let $f: (\mathbb{C}^3, 0) \to (\mathbb{C}, 0)$ be a germ of a holomorphic function at an isolated critical point. A transversal to the equivalence class of this germ in the space of all function germs is called a *versal deformation* of f. Instead of giving this informal definition a rigorous meaning (see [4]), we shall show how to construct a *mini*versal deformation (that is, a transversal of minimal possible dimension).

Let $\mathbb{C}\{x, y, z\}$ be the algebra of all germs. The tangent space to the orbit of a germ f under the action of the group of diffeomorphisms is just the gradient ideal $(f_x, f_y, f_z) \subset \mathbb{C}\{x, y, z\}$. The quotient space

$$Q = \mathbb{C}\{x, y, z\}/(f_x, f_y, f_z)$$

is called the *local algebra* of f. Let $e_1, \ldots, e_\mu \in \mathbb{C}\{x, y, z\}$ be representatives of a basis in Q. Then

$$F(x, y, z, \lambda) = f + \lambda_1 e_1 + \cdots + \lambda_\mu e_\mu$$

is a miniversal deformation. The number of parameters $\mu = \dim_\mathbb{C} Q$ of a miniversal deformation is called the *multiplicity* or the *Milnor number* of the critical point; it is equal to the number of nondegenerate critical points into which it bifurcates under a deformation of general position. (The local algebra can be considered as the algebra of functions over μ merged critical points.)

EXAMPLE. Let $f = x^{\mu+1} - yz$ be a germ of type A_μ. Then $Q \simeq \mathbb{C}[x]/(x^\mu)$ and
$$x^{\mu+1} - yz + \lambda_1 x^{\mu-1} + \cdots + \lambda_\mu$$
is a miniversal deformation of f.

The hypersurface Δ in the parameter space of a miniversal deformation whose points correspond to the functions $F_\lambda = F(\cdot, \lambda)$ whose zero level $V_\lambda = F_\lambda^{-1}(0)$ is singular, is called the *discriminant of* f.

EXAMPLE. The discriminant of the germ A_μ is the discriminant of the space of polynomials
$$x^{\mu+1} + \lambda_1 x^{\mu-1} + \cdots + \lambda_\mu.$$

THEOREM. *The discriminant of a simple singularity of type A_μ, D_μ, E_μ is isomorphic to the discriminant of the Weyl group and of the simple Lie algebra of the same type.*

The original proof of this theorem, due to Brieskorn [22], is based on a relationship between simple singularities and simple Lie algebras to be described in § 3.2. In § 2.5 we shall outline a simpler proof, due to Looijenga [26].

2.4. **Monodromy group** [5]. We now describe a map

simple singularities   ⟶   Weyl groups
∩   ∩
isolated critical points   ⟶   reflection groups.

Given a miniversal deformation F of a germ $f: (\mathbb{C}^3, 0) \to (\mathbb{C}, 0)$, we consider the family of surfaces ([2]) $V_\lambda = F_\lambda^{-1}(0)$, $\lambda \in \mathbb{C}^\mu$. This family is a locally trivial bundle outside the discriminant $\Delta \subset \mathbb{C}_\mu$. Associated with it, we have the homology bundle $H_*(V_\lambda)$ over $\mathbb{C}^\mu \setminus \Delta$, which is equipped with the flat *Gauss-Manin connection*: cycles can be transported to nearby fibers by continuity. The holonomy of this connection determines the *monodromy representation*
$$\pi_1(\mathbb{C}^\mu \setminus \Delta) \to \operatorname{Aut} H_*(V_{\lambda_0})$$

([2]) More precisely, fix a ball $B \subset \mathbb{C}^3$ around the origin with a radius so small that its boundary is transversal to $F_0^{-1}(0)$. Then fix a neighborhood of the origin in the parameter space, over which the fibers $F_\lambda^{-1}(0)$ are still transversal to the boundary of B. Our family is defined over this neighborhood; it consists of the surfaces $V_\lambda = F_\lambda^{-1}(0) \cap B$. In what follows we shall omit these details, all the more so as in our examples the family $F_\lambda^{-1}(0)$ will be defined globally in $\mathbb{C}^3 \times \mathbb{C}^\mu$.

of the fundamental group of the complement to the discriminant in the homology space of a nonsingular fiber over a distinguished point λ_0 (which must not lie in the discriminant).

EXAMPLE. In the family P_λ of polynomials of degree $\mu + 1$ in one variable, the monodromy representation in $H_0(P_\lambda^{-1}(0)) \simeq \mathbb{Z}^{\mu+1}$ is defined by the standard epimorphism of the braid group of the $(\mu + 1)$th string onto a permutation group.

EXERCISE. Find the monodromy for a miniversal family of surfaces of type $A_1: x^2 + y^2 + z^2 = \lambda$.

MILNOR'S THEOREM [7]. *A nonsingular fiber V_λ is homotopy equivalent to the union of μ spheres of middle dimension ($= \dim_\mathbb{C} V_\lambda = 2$).*

COROLLARY. $H_*(V_\lambda) = H_0(V_\lambda) \oplus H_2(V_\lambda) = \mathbb{Z} \oplus \mathbb{Z}^\mu$.

The image W of the monodromy representation in $\operatorname{Aut}(H_2(V_\lambda))$ is called the *monodromy group* of the germ f.

In the middle homology space $H_2(V_\lambda)$ of a nonsingular fiber the intersection number of the cycles defines an integral bilinear form $\langle\,,\,\rangle$. This form is symmetric (since $\dim_\mathbb{C} V_\lambda = 2$ is even), but not necessarily positive or negative definite (it may also be degenerate).

EXERCISE. Prove that the self-intersection number of the real sphere $x^2 + y^2 + z^2 = 1$ in the complex surface with the same equation is -2.

THEOREM [5]. *The monodromy group is generated by reflections in the hyperplanes that are orthogonal to vectors of some basis of the lattice $H_2(V_\lambda)$ with respect to the intersection form $\langle\,,\,\rangle$.*

THEOREM [5]. *The intersection form is definite (it is, in fact, negative definite) if and only if the germ f has a simple singularity A_μ, D_μ, or E_μ. In this case the pair $(W, \langle\,,\,\rangle)$ is isomorphic to the Weyl group of the same type.*

EXERCISE (proof of Milnor's Theorem). (1) The Morse index of the squared modulus of a holomorphic function of n variables at a nondegenerate critical point with nonzero critical value is n.

(2) Let $(f_1, \ldots, f_n)$ be a regular sequence of germs of holomorphic functions $(\mathbb{C}^n, 0) \to (\mathbb{C}, 0)$, $(\tilde{f}_1, \ldots, \tilde{f}_n)$ a generic perturbation of the sequence, $V^p = \tilde{f}_1^{-1}(0) \cap \cdots \cap \tilde{f}_{n-p}^{-1}(0)$ a local nonsingular p-dimensional complete intersection, $V^1 \subset \cdots \subset V^n = \mathbb{C}^n$. Prove (using Morse theory for the function $|\tilde{f}_{n-p}|^2$ restricted to V^{p+1}) that the space V^{p+1}/V^p is homotopy equivalent to a bouquet of $(p+1)$-dimensional spheres.

(3) If a finite cell complex is contractible, each of its p-skeletons is homotopy equivalent to a bouquet of p-dimensional spheres.

(4) Deduce from this that V^p is homotopically equivalent to the bouquet of p-dimensional spheres.

(5) Show that when $p = n - 1$, the number of these spheres is equal to the number of critical points of $\tilde{f}_1$.

2.5. Period map [5, 26].
We now describe a (multi-valued) map of the complement of the discriminant of a miniversal deformation F_λ of a germ f to the cohomology space $H^2(V_{\lambda_0}, \mathbb{C})$ of the distinguished fiber. This enables us to prove the theorem in §2.3 about discriminants of simple singularities.

Fix a "complex volume" $dx \wedge dy \wedge dz$ in $\mathbb{C}^3$ and consider the 2-form $\omega = dx \wedge dy \wedge dz/dF_\lambda$. This expression can be understood as the holomorphic 2-form which, when multiplied by dF_λ, becomes the volume form. Such a 2-form is determined uniquely up to addition of a 2-form which is divisible by dF_λ. Hence its restriction to a nonsingular fiber $V_\lambda = F_\lambda^{-1}(0)$ is uniquely determined and is a holomorphic 2-form there.

The *period map*

$$\lambda \in \mathbb{C}^\mu \setminus \Delta \longmapsto [\omega] \in H^2(V_\lambda, \mathbb{C})$$

is a holomorphic section of the cohomology bundle.

Move the cohomology class $[\omega]$ to the fiber over the distinguished point using the Gauss-Manin connection. The result depends on the path and is uniquely determined up to a monodromy transformation. Thus, only the map into the set of orbits is well defined:

$$\mathbb{C}^\mu \setminus \Delta \xrightarrow{[\omega]} H^2(V_{\lambda_0}, \mathbb{C})/W.$$

In the case of simple singularities, W acts in $H^2(V_{\lambda_0}, \mathbb{C})$ as a Weyl group, orbits form a μ-dimensional complex space, and our period map is holomorphic.

THEOREM (Looijenga). *In the case of simple singularities the period map $[\omega]$ extends holomorphically to the discriminant as an isomorphism*

$$\mathbb{C}^\mu \xrightarrow{[\omega]} H^2(V_{\lambda_0}, \mathbb{C})/W,$$

which maps the discriminant Δ onto the manifold of singular orbits of the monodromy group.

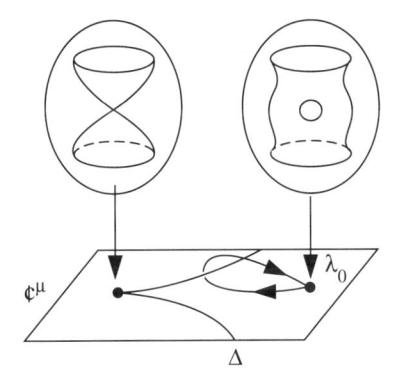

§3. Resolution of singularities

3.1. Springer's construction [21]. Consider the projection $\mathfrak{g} \to \mathfrak{g}/G$ of a complex simple Lie algebra to the orbit space of the adjoint representation. This projection defines a bundle over the complement of the discriminant, whose fiber is a regular orbit X_λ. We want to study the monodromy representation

$$\pi_1((\mathfrak{g}/G) \setminus \Delta) \to \operatorname{Aut} H_*(X_\lambda),$$

generated by the Gauss-Manin connection in this bundle.

EXAMPLE. sl_2 is the 3-dimensional algebra of matrices $\begin{pmatrix} a & b \\ c & -a \end{pmatrix}$. The projection $\mathfrak{g} \to \mathfrak{g}/G : (a, b, c) \mapsto a^2 + bc$ is a miniversal deformation of a quadratic cone with a singularity of type A_1. The monodromy group is isomorphic to $\mathbb{Z}_2$ and acts on $H_*(X_\lambda) = H_0 \oplus H_2 \simeq \mathbb{Z} \oplus \mathbb{Z}$, trivially on the first summand and changing sign on the second.

It is known that $H_*(X_\lambda) \simeq \mathbb{Z}^{|W|}$ and that the monodromy group is the regular representation of the Weyl group W.([3]) In fact, the fiber X_λ is the total space of a vector bundle over the flag manifold F of $\mathfrak{g}$, and F decomposes into $|W|$ complex Schubert cells. The closures of the Schubert cells form a basis of cycles in $H_*(F)$ (in particular, $H_*(F) = H_{\text{even}}(F)$).

EXAMPLE. $\mathfrak{g} = sl_{\mu+1}$. A regular orbit consists of matrices conjugate to a diagonal matrix with simple spectrum. Therefore $X_\lambda = SL_{\mu+1}/T$ is the quotient space of the group by the stabilizer of this diagonal matrix, i.e., by the maximal torus $T = \{\operatorname{diag}(x_0, \ldots, x_\mu), x_0 \cdots x_\mu = 1\}$. The group P of upper-triangular unipotent matrices acts freely on X_λ and defines a fibration $X_\lambda \xrightarrow{P} X_\lambda/P$ with a contractible fiber. The space X_λ/P is isomorphic to $SL_{\mu+1}/T \cdot P$—the quotient space of the general matrix group by *all* upper-triangular matrices. The latter group is the stabilizer of the coordinate flag $\mathbb{C}^1 \subset \mathbb{C}^2 \subset \cdots \subset \mathbb{C}^{\mu+1}$, so that X_λ/P is a flag manifold. A Schubert cell consists of flags whose intersections with the coordinate flag have fixed dimensions. It contains just one flag spanned by the same basis vectors, perhaps ordered differently. Hence the number of Schubert cells is $(\mu + 1)!$.

Springer's construction, which will be described below (in a geometrical form due to Slodowy), singles out each irreducible representation of the Weyl group exactly once from the regular monodromy representation and determines a correspondence between such representations and nilpotent orbits.

Let $\mathfrak{g}$ be an algebra of type A_μ, D_μ, or E_μ. The fiber X_0 of the projection $\mathfrak{g} \to \mathfrak{g}/G$ is the cone of nilpotent elements in G, a singular variety of codimension μ, stratified by nilpotent orbits.

EXAMPLE. The nilpotent orbits in $sl_{\mu+1}$ are indexed by the orders $k_1 \geq$

([3]) Recall that the regular representation is the action of a group in the space of functions on the group by shifts of the argument. The number of times any irreducible representation is contained in the regular representation is equal to the dimension of that irreducible representation.

$k_2 \geq \cdots$ of the Jordan blocks and therefore by Young diagrams with $\mu + 1$ cells.

Let Y_d be a nilpotent orbit of codimension d in X_0. Specialize the family X_λ near Y_d. In other words, choose a germ of a transversal to Y_d in $\mathfrak{g}$ at some point (this transversal has dimension $\mu + d$) and restrict the projection onto $\mathfrak{g}/G$ to this germ. We obtain a family

$$(\mathbb{C}^{\mu+d}, 0) \xrightarrow{V_\lambda} (\mathfrak{g}/G, 0)$$

of μ-dimensional varieties, which is a locally trivial bundle outside the discriminant. The fiber $V_0 = X_0 \cap (\mathbb{C}^{\mu+d}, 0)$ is the intersection of the cone of nilpotents with the transversal, the typical fiber V_λ is the nonsingular restriction of a regular orbit to the transversal.

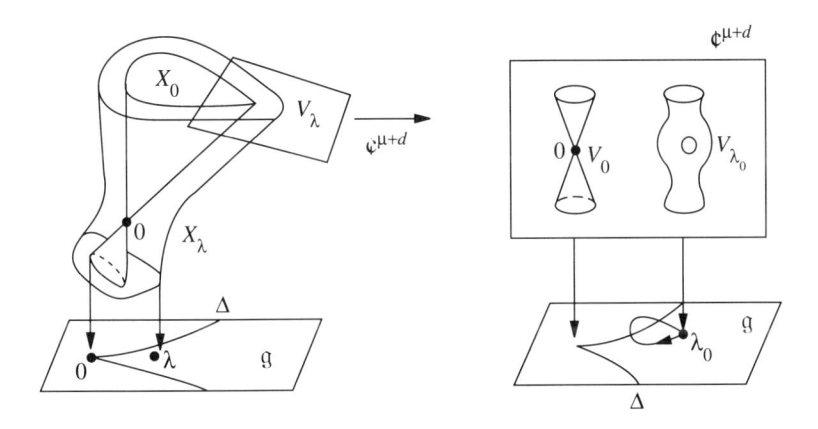

THEOREM. *The monodromy representation*

$$\pi_1((\mathfrak{g}/G) \setminus \Delta) \to \mathrm{Aut}\; H_d(V_{\lambda_0})$$

in the middle homology of the distinguished nonsingular fiber $V_{\lambda_0}^d$ is an irreducible representation of the Weyl group. Each irreducible representation of the Weyl group can be obtained by this construction exactly once.

EXAMPLE. $d = 0$. A transversal to a regular nilpotent orbit is μ-dimensional and projects into $\mathfrak{g}/G$ isomorphically. Hence V_λ is a point and the monodromy representation is the identity representation.

3.2. Brieskorn's theorem [21]. This theorem gives a correspondence

Lie algebras $A_\mu, D_\mu, E_\mu \longrightarrow$ simple surface singularities.

Let Y_d be a subregular nilpotent orbit in $\mathfrak{g}$, that is, a nilpotent orbit of next highest dimension after a regular one. In that case $d = 2$,([4]) so that V_λ is

---

([4]) And not 1, as one might think. The point is that the adjoint representation of a semisimple Lie algebra is isomorphic to the co-adjoint representation, since the Killing form is nondegenerate, and so the orbits are symplectic manifolds.

a family of surfaces and the singularity of the surface V_0 is isolated.

THEOREM [22]. *The surface V_0 cut out on the nilpotent cone by the transversal $(\mathbb{C}^{\mu+2}, 0)$ to a subregular nilpotent orbit has a simple singularity A_μ, D_μ, or E_μ at zero, of the same type as the algebra, and the family*

$$(\mathbb{C}^{\mu+2}, 0) \xrightarrow{V_\lambda} (\mathfrak{g}/G, 0)$$

is a miniversal deformation of this singularity.

COROLLARY 1. *Springer's construction, applied to a subregular nilpotent orbit, yields the Coxeter representation of the Weyl group.*

COROLLARY 2. *The period map $[dx \wedge dy \wedge dz/dF_\lambda]$ (see §2.5) yields an isomorphism $\mathfrak{g}/G \to H^2(V_\lambda, \mathbb{C})/W$.*

Indeed, it is known that Kirillov's symplectic structure on orbits defines an isomorphism $\mathfrak{g}/G \to H^2(F, \mathbb{C})/W$. It can be shown that $dx \wedge dy \wedge dz/dF_\lambda$ is the restriction of Kirillov's structure to the intersections of orbits with a transversal, when they are identified with fibers of a versal deformation as in Brieskorn's theorem.

EXAMPLE. $sl_{\mu+1}$. In that case a subregular nilpotent matrix has two Jordan blocks of orders μ and 1. As a transversal to its orbit one can take the family of matrices

$$M = \begin{bmatrix} \begin{array}{cccc|c} 0 & -1 & & 0 & 0 \\ & 0 & -1 & & \\ & & \ddots & \ddots & \vdots \\ & & & 0 & -1 \\ 0 & & & 0 & 0 \\ \hline 0 & \cdots & & 0 & 0 \end{array} \end{bmatrix} + \begin{bmatrix} \begin{array}{cccc|c} -x & & & & 0 \\ & -x & & 0 & \\ & & \ddots & & \vdots \\ & & & -x & 0 \\ \lambda_{\mu-1} & \cdots & \lambda_1 & -x & y \\ \hline z & 0 & \cdots & 0 & \mu x \end{array} \end{bmatrix}.$$

The characteristic polynomial

$$\det(M - \lambda) = \lambda^{\mu+1} + \Delta_1 \lambda^{\mu-1} + \cdots + \Delta_\mu$$

has the form

$$-yz - (\mu x + \lambda)[(x-\lambda)^\mu + \lambda_1(x-\lambda)^{\mu-2} + \cdots + \lambda_{\mu-2}(x-\lambda) + \lambda_{\mu-1}].$$

The parameter transformation $(\lambda_1, \ldots, \lambda_{\mu-1}) \mapsto (\Delta_1, \ldots, \Delta_{\mu-1})$ is a diffeomorphism, and the relation

$$yz + \mu(x^{\mu+1} + \lambda_1 x^{\mu-1} + \cdots + \lambda_{\mu-1} x) + \Delta_\mu = 0$$

is a miniversal deformation of the simple type A_μ singularity with the equation $yz + \mu x^{\mu+1} = 0$.

3.3. Simultaneous resolution [22]. The following construction of Grothendieck gives a simultaneous resolution of the singularities of the fibers of the projection $\mathfrak{g} \to \mathfrak{g}/G$, and, at the same time, of the surfaces of a miniversal family of the corresponding simple singularity. We shall describe it for the example $\mathfrak{g} = sl_{\mu+1}$.

Let $X \subset F \times \mathfrak{g}$ be the set of pairs (f, x) in the product of the flag manifold and the corresponding simple Lie algebra, such that x transforms f into itself. Projection of X onto the first factor shows that X is not singular: the fiber over f is the stationary subalgebra of f, i.e., the triangular subalgebra in the coordinate system associated with the flag. Assign to a pair $(f, x) \in X$ a sequence of numbers $(x_0, \ldots, x_\mu)$, $\sum x_i = 0$, defined as follows: The matrix x that preserves the flag $0 \subset \mathbb{C}^1 \subset \mathbb{C}^2 \subset \cdots \subset \mathbb{C}^{\mu+1}$ acts on the successive quotients $\mathbb{C}^{k+1}/\mathbb{C}^k$ through multiplication by $x_0, \ldots, x_\mu$ respectively. We obtain a projection $X \to \mathbb{C}^\mu$ that assigns to each pair (f, x) the sequence of eigenvalues of x, ordered by the flag f.

The fiber of the projection $X \to \mathfrak{g}$ onto the second factor consists of all flags that are invariant with respect to a given matrix $x \in \mathfrak{g}$. For a regular matrix it consists of $|W| = (\mu + 1)!$ points—the coordinate flags in an eigenbasis. In the case of a multiple spectrum, the fiber may have positive dimension. We have a commutative diagram:

$$\begin{array}{ccc} X & \longrightarrow & \mathfrak{g} \\ \downarrow & & \downarrow \\ \mathbb{C}^\mu & \longrightarrow & \mathbb{C}^\mu/W = \mathfrak{g}/G. \end{array}$$

THEOREM (Grothendieck). *The map $X \to \mathfrak{g}$ gives a simultaneous resolution of singularities of the fibers of the projection $\mathfrak{g} \to \mathfrak{g}/G$.*

In other words, the fibers $X \to \mathbb{C}^\mu$ are nonsingular and they are resolutions of the corresponding fibers of $\mathfrak{g} \to \mathfrak{g}/G$. This simultaneous resolution of the singularities of all the fibers becomes possible after one goes over to a W-covering $\mathbb{C}^\mu \to \mathfrak{g}/G$ which is ramified over the discriminant. Indeed, the bundle $X \xrightarrow{X_{\bar{\lambda}}} \mathbb{C}^\mu$ over such a covering can be topologically trivialized; this explains why the monodromy representation $\pi_1((\mathbb{C}^\mu/W) \setminus \Delta) \to \operatorname{Aut} H_*(X_{\bar{\lambda}})$ of the braid group factors through the Weyl group.

COROLLARY (Brieskorn). *Miniversal deformations of simple surface singularities have a simultaneous resolution.*

One need only to replace $\mathfrak{g}$ in the above diagram by a transversal to a subregular nilpotent orbit and X by the inverse image of the transversal.

EXAMPLE. sl_2. In this case the flag manifold is isomorphic to the projective line $\mathbb{C}P^1$ of one-dimensional subspaces in $\mathbb{C}^2$. The nilpotent cone $\det\begin{pmatrix} a & b \\ c & -a \end{pmatrix} = 0$ is quadratic. Its inverse image in $\mathbb{C}P^1 \times \mathfrak{g}$ consists of pairs (nilpotent matrix, invariant line of the matrix). The line is unique if the nilpotent matrix is not zero. The inverse image of zero is the entire flag manifold $\mathbb{C}P^1$. This resolution of the cone in $\mathbb{C}^3$ is achieved by blowing up (σ-process) at zero. The complement of the quadratic cone is doubly covered by the map $X \to \mathfrak{g}$.

3.4. **Exceptional divisor** [22, 30]. In the resolution $\tilde{V} \xrightarrow{\pi} (V, 0)$ of a singular point an *exceptional divisor* is glued onto a complex surface. This defines a *resolution diagram*. The vertices of the diagram correspond to the irreducible components $D = \bigcup D_i$. Vertices i, j are connected by k edges, where k is the multiplicity of the intersection $D_i \cap D_j$. We shall now find the diagrams of minimal resolutions of simple singularities of type A.

EXAMPLE. A_2. To a subregular nilpotent $x \in sl_3$ we associate the flag $\mathbb{C}^1 \subset \mathbb{C}^2 \subset \mathbb{C}^3$: $\mathbb{C}^2 = \ker x$, $\mathbb{C}^1 = \operatorname{im} x$. An invariant flag of the operator x is either the same plane $\mathbb{C}^2$ and an arbitrary line in it, or the line $\mathbb{C}^1$ and an arbitrary plane containing it. The exceptional divisor therefore consists of two projective straight lines intersecting at a point.

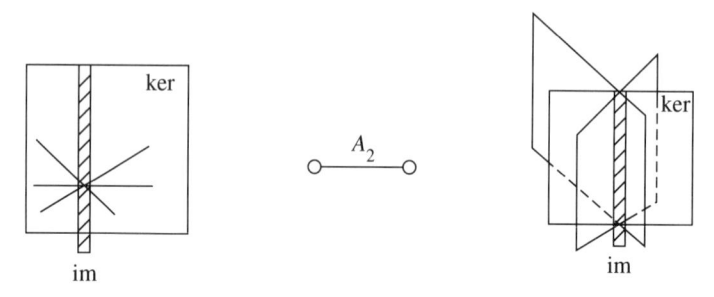

The situation is similar in the case of A_μ. The invariant flags of a subregular nilpotent $x \in sl_{\mu+1}$ are described as follows. Let $K^d = \ker x^{d-1}$, $I^d = \operatorname{im} x^{\mu-d}$. Then $I^1 \subset \cdots \subset I^{k-1} \subset L^k \subset K^{k+1} \subset \cdots \subset K^\mu \subset \mathbb{C}^{\mu+1}$ is an invariant flag, where L is determined by a straight line in the plane K^{k+1}/I^{k-1}, $k = 1, \ldots, \mu$. Hence the exceptional divisor consists of μ projective lines $D_1, \ldots, D_\mu$. When $|i - j| > 1$, D_i and D_j do not intersect, while the intersection $D_k \cap D_{k+1}$ consists of one point (this is evident if one applies the arguments of the above example to the three-dimensional space K^{k+2}/I^{k-1}). Thus the resolution diagram is the Dynkin diagram A_μ.

THEOREM (Artin). *The minimal resolution diagram of a simple singular point of type A_μ, D_μ, or E_μ is the Dynkin diagram of the same type.*

In fact, the exceptional divisor on a minimal resolution $\tilde{V}_0 \to V_0$ consists of μ projective lines, which represent a basis of cycles in $H_2(\tilde{V}_0)$. The intersection form in this basis is as prescribed by the Dynkin diagram: $\langle D_i, D_j \rangle = 0$ or 1, $i \neq j$, $\langle D_i, D_i \rangle = -2$.

§4. Parabolic singularities

4.1. **Unimodal critical points** [20, 3]. In §1.3 we listed the normal forms A, D, E of simple singularities of functions in three variables, that is, discrete equivalence classes of germs of holomorphic functions. We will now extend our classification list to include unimodal singularities, that is, classes that depend on one continuous invariant.

THEOREM (Arnol'd). *A unimodal germ* $f : (\mathbb{C}^3, 0) \to (\mathbb{C}, 0)$ *of a holomorphic function is equivalent to one of the following* (*here a is the modulus*):
(a) *parabolic singularities*:

$$P_8: x^3 + y^3 + z^3 + axyz, \qquad a^3 + 27 \neq 0,$$
$$X_9: x^4 + y^4 + z^2 + ax^2y^2, \qquad a^2 \neq 4,$$
$$J_{10}: x^3 + y^6 + z^2 + ax^2y^2, \qquad 4a^3 + 27 \neq 0;$$

(b) *hyperbolic singularities* $(1/p + 1/q + 1/r < 1)$:

$$T_{p,q,r}: x^p + y^q + z^r + axyz, \qquad a \neq 0;$$

(c) *fourteen exceptional singularities*:

$$K_{12}: x^3 + y^7 + z^2 + axy^5, \qquad W_{13}: x^4 + xy^4 + z^2 + ay^8,$$
$$K_{13}: x^3 + xy^5 + z^2 + ay^8, \qquad Q_{10}: x^3 + y^4 + yz^2 + axy^3,$$
$$K_{14}: x^3 + y^8 + z^2 + axy^6, \qquad Q_{11}: x^3 + y^2z + xz^3 + az^5,$$
$$Z_{11}: x^3y + y^5 + z^2 + axy^4, \qquad Q_{12}: x^3 + y^5 + yz^2 + axy^4,$$
$$Z_{12}: x^3y + xy^4 + z^2 + ax^2y^3, \qquad S_{11}: x^4 + y^2z + xz^2 + ax^3z,$$
$$Z_{13}: x^3y + y^6 + z^2 + axy^5, \qquad S_{12}: x^2y + y^2z + xz^3 + az^5,$$
$$W_{12}: x^4 + y^5 + z^2 + ax^2y^3, \qquad U_{12}: x^3 + y^3 + z^4 + axyz^2.$$

The normal forms of unimodal germs of a greater number of variables are obtained from those listed above by adding a nondegenerate quadratic form in the remaining variables.

Some of the above normal forms are quasihomogeneous. Recall that a polynomial f is *quasihomogeneous* if one can assign to its variables $(x_1, \ldots, x_n)$ positive weights $(\alpha_1, \ldots, \alpha_n)$ in such a way that f becomes homogeneous (of degree 1):

$$f(t^{\alpha_1} x_1, \ldots, t^{\alpha_n} x_n) = t f(x_1, \ldots, x_n).$$

By the Euler formula, a quasihomogeneous germ lies in its gradient ideal:

$$f = \alpha_1 x_1 \frac{\partial f}{\partial x_1} + \cdots + \alpha_n x_n \frac{\partial f}{\partial x_n}.$$

This property is invariant under changes of variables and characterizes the entire equivalence class of the germ.[5] The quasihomogeneous classes among those listed are simple classes A_μ, D_μ, E_μ, all parabolic classes P_8, X_9, J_{10}, and also exceptional classes with $a = 0$.

In this and in the next sections we shall examine the quasihomogeneous unimodal singularities and generalize to them the description of zero level surfaces and discriminants of simple singularities given in §1.5 and §2.

---

[5] By a theorem of Saito [24], a germ of a holomorphic function at an isolated critical point representing the zero class in its local algebra is quasihomogeneous in suitable local coordinates.

4.2. Elliptic curves [25].

Let E be an elliptic curve, $L \xrightarrow{\mathbb{C}} E$ a line (= holomorphic one-dimensional vector) bundle over E. Let $\hat{L}$ denote the surface obtained from L by one-point compactification (i.e., addition of a common point at infinity ∞ to all fibers) and removal of the zero section. $\hat{L}$ is a surface with one singular point at ∞.

More precisely, in order to define $\hat{L}$ as an algebraic surface, we describe its algebra $\mathscr{A}$ of regular functions. Let L^k be the tensor power of L, $k = 0, 1, 2, \ldots$. The canonical isomorphism $L^k \otimes L^e = L^{k+e}$ defines the multiplication of holomorphic sections:

$$\Gamma(L^k) \otimes \Gamma(L^e) \to \Gamma(L^{k+e}).$$

Now put

$$\mathscr{A} = \bigoplus_{k=0}^{\infty} \Gamma(L^k).$$

Let $c(L)$ be the self-intersection number of the zero section of L. The algebra $\mathscr{A}$ separates points on $\hat{L}$ if and only if $c(L) > 0$. Thus, we obtain a sequence [6] $\hat{L}_c$, $c = 1, 2, 3, \ldots$, of singular algebraic surfaces.

EXERCISE.
(1) $c(L^k) = kc(L)$.
(2) If $c(L) < 0$, then $\mathscr{A} \simeq \mathbb{C}$.
(3) Prove that if $c(L) = 0$, then $\mathscr{A} \simeq \mathbb{C}$ or $\mathbb{C}[x]$.
(4) If $c(L) > 0$, then $\dim \Gamma(L) = c(L)$.

THEOREM (Saito [25]). *The surface $\hat{L}_c$ is imbedded in $\mathbb{C}^3$ if and only if $c = 3, 2, 1$. Its image in $\mathbb{C}^3$ has a parabolic singular point of type P_8, X_9, J_{10} respectively. All parabolic surface singularities are obtained by this construction for an appropriate choice of the elliptic curve E.*

COROLLARY. *As a continuous invariant of a parabolic singularity one can take the invariant of the corresponding elliptic curve.*

REMARK. The moduli space of elliptic curves $E = \mathbb{C}/\mathbb{Z} \oplus \tau \mathbb{Z}$ is the quotient space of the upper half-plane $\{\tau | \operatorname{Im} \tau > 0\}$ with respect to the action of the group $\operatorname{PSL}_2(\mathbb{Z})$ of integral linear-fractional transformations $\tau \mapsto (a\tau + b)/(c\tau + d)$ and is isomorphic to $\mathbb{C}^1$.

EXAMPLE. The surface $P_8 : x^3 + y^3 + z^3 + axyz = 0$ is a Hopf cone over a cubic in $\mathbb{C}P^2$; hence it can obviously be identified with $\hat{L}_3$.

EXERCISE. (1) Prove that if $c \geq 3$, then the number of generators of the algebra $\mathscr{A}$ is not less than c.

---

[6] In the category of holomorphic bundles the equivalence class also depends on a continuous parameter—a point on the "Jacobian" of the curve, in this case—on the elliptic curve E itself. However, when $c(L) \neq 0$, nonequivalent bundles are transformed into each other under the action of holomorphic automorphisms (shifts) of an elliptic curve. Therefore the holomorphy type of the *space* of a bundle is uniquely determined by the number $c \neq 0$.

(2) For $c = 1, 2, 3$ find the degrees of the generators x, y, z of the graded algebra $\mathscr{A} = \bigoplus \Gamma(L^k)$ and the degree of the relation $f(x, y, z) = 0$ among them. Compare this result with the quasihomogeneous weights of the parabolic singularities.

(3) Prove that a nonsingular intersection of two quadrics in $\mathbb{C}P^3$ is an elliptic curve.

(4) Prove that $\hat{L}_4$ is embedded in $\mathbb{C}_4$ as the complete intersection (notation: Q_7) of two nondegenerate quadratic cones.

REMARK. The inverse quasihomogeneous weights $(2, 2, 2, 2)$, $(3, 3, 3)$, $(2, 4, 4)$, $(2, 3, 6)$ of the singularities Q_7, P_8, X_9, J_{10} coincide with the "ramification numbers" of the groups of motions of the Euclidean plane generated by reflections in the sides of a square and of the three special triangles with angles $\left(\frac{\pi}{3}, \frac{\pi}{3}, \frac{\pi}{3}\right)$, $\left(\frac{\pi}{2}, \frac{\pi}{4}, \frac{\pi}{4}\right)$, $\left(\frac{\pi}{2}, \frac{\pi}{3}, \frac{\pi}{6}\right)$.

4.3. Complex crystallographic groups [8].

The complex crystallographic groups are classified by extended Dynkin diagrams. In the cases A_μ, D_μ, E_μ they can be described as follows.

A Weyl group W acts by matrices with integer entries on the space $\mathbb{C}^\mu$, which is the complexification of the lattice $\mathbb{Z}^\mu$. This action maps the lattice $\mathbb{Z}^\mu \oplus \tau \mathbb{Z}^\mu$ into itself, where τ is a point in the upper half-plane. A complex crystallographic group is a semidirect product, namely, an extension of W by this lattice:

$$0 \to \mathbb{Z}^\mu \oplus \tau \mathbb{Z}^\mu \to \tilde{W} \to W \to 1.$$

In fact, the lattice $\mathbb{Z}^\mu \oplus \tau \mathbb{Z}^\mu$ is invariant under integral fractional-linear transformations of τ, so that we obtain a family of actions of $\tilde{W}$ in $\mathbb{C}^\mu$ parametrized by equivalence classes of elliptic curves.

The Chevalley Theorem for complex crystallographic groups, proved by Bernstein and Shvartsman, describes the orbit space $\mathbb{C}^\mu / \tilde{W}$. The factorization may be accomplished in two steps. The space $\mathbb{C}^\mu / \mathbb{Z}^\mu \oplus \tau \mathbb{Z}^\mu$ is isomorphic to the product $E_\tau \times \cdots \times E_\tau$ of μ copies of the elliptic curve $E_\tau = \mathbb{C} / \mathbb{Z} \oplus \tau \mathbb{Z}$ on which the Weyl group W acts. Thus, $\mathbb{C}^\mu / \tilde{W} = E_\tau^\mu / W$.

By a *quasihomogeneous projective space* $\mathbb{C}P_\alpha^\mu$ we mean the space of complex orbits in $\mathbb{C}^{\mu+1} \setminus 0$ of the one-parameter group generated by the Euler field $\sum_{i=0}^\mu \alpha_i z_i \partial / \partial z_i$ with positive integer weights $\alpha = (\alpha_0, \ldots, \alpha_\mu)$. The space $\mathbb{C}P_\alpha^\mu$ is a compact algebraic variety, in general singular, which is isomorphic to the quotient space of $\mathbb{C}P^\mu$ by the action of the finite group $\mathbb{Z}_{\alpha_0} \times \cdots \times \mathbb{Z}_{\alpha_\mu}$.

THEOREM (Bernstein-Shvartsman). *The orbit space $\mathbb{C}^\mu / \tilde{W}$ of a complex crystallographic Coxeter group is a quasihomogeneous projective space $\mathbb{C}P_\alpha^\mu$. The tuple α of quasihomogeneous weights in cases $W = A_\mu$, D_μ, or E_μ coincides with the tuple $(d_0, \ldots, d_\mu)$ of weights of the extended Dynkin diagram (see §1.6), that is, of dimensions of irreducible representations of the finite quaternion group of the same type.*

The *discriminant* of a complex crystallographic group is the hypersurface Δ_τ of nonregular orbits in $\mathbb{C}P_\alpha^\mu = E_\tau^\mu/W$. Unlike the orbit space itself, it depends on the invariant τ of the elliptic curve.

EXAMPLES. A_1. The group A_1 acts on the elliptic curve $E_\tau = \mathbb{C}/\mathbb{Z} \oplus \tau\mathbb{Z}$ (or $y^2 = x(x-1)(x-2)(x-\lambda(\tau))$) by the reflection $z \to -z$ (or $(x, y) \to (x, -y)$). Thus the orbit space consists of four ramification points $(0, 1, 2, \lambda)$, namely, of the elements of second order in the group E_τ.

A_2. The Weyl group A_2 acts by permutations of factors in the product $E_\tau \times E_\tau \times E_\tau$, leaving invariant the "plane" with coordinate sum zero. Recalling the definition of the sum of points on a cubic $E_\tau \subset \mathbb{C}P^2$, we see that three points $e_1, e_2, e_3 \in E_\tau$ such that $e_1 + e_2 + e_3 = 0$ are exactly the three points at which E_τ intersects a line in $\mathbb{C}P^2$. The line itself determines its points of intersection with E_τ up to a permutation. Therefore, $\mathbb{C}^2/\tilde{A}_2$ is the dual projective plane $\mathbb{C}P^{2*}$ of all lines in $\mathbb{C}P^2$. Nonregular orbits are lines tangent to E_τ in $\mathbb{C}P^2$. Hence the discriminant is a curve (of degree 6), which is projectively dual to the cubic. It has nine cusp points, only three or one of which are real. These nine points correspond to the inflection points of the cubic, namely, to the nine points of order three in E_τ.

EXERCISE. (1) Prove that the discriminant $\tilde{A}_3$ is the projectively dual surface to a nonsingular intersection of two quadrics in $\mathbb{C}P^3$.

(2) Prove the Bernstein-Shvartsman Theorem for the sequence A_μ, by identifying the orbit space E_τ^μ/W with the projectivized space of holomorphic sections of the line bundle $L \to E_\tau$ of degree $c(L) = \mu + 1$. Show that the discriminant consists of the sections with multiple zeroes.

4.4. Structure of the discriminant [27, 28]. Let us examine a miniversal deformation of a parabolic singularity, taking as an example $P_8 : x^3 + y^3 + z^3 + axyz$. The monomials $(1, x, y, z, xy, yz, zx, xyz)$ form a basis of the local algebra $Q = \mathbb{C}[x, y, z]/(x^2, y^2, z^2)$ for $a = 0$; we may assume that this is also true for all admissible a ($a^3 + 27 \neq 0$). A miniversal deformation

$$x^3 + y^3 + z^3 + axyz + \lambda_1 xy + \lambda_2 yz + \lambda_3 zx + \lambda_4 x + \lambda_5 y + \lambda_6 z + \lambda$$

is a quasihomogeneous polynomial (necessarily $\deg x = \deg y = \deg z = 1$). The invariant a of the parabolic singularity is one of the deformation parameters and has quasihomogeneous degree 0. The tuple of weights $\alpha = (1, 1, 1, 2, 2, 2, 3)$ of the remaining parameters is positive and coincides, as can be seen, with the tuple of weights of the extended Dynkin diagram E_6 (see §1.6). The discriminant of the versal deformation is a quasihomogeneous hypersurface in the parameter space. The quasihomogeneous projectivization of the discriminant on a hypersurface $a = $ const in this space defines a hypersurface ∇_a in the quasihomogeneous projective space $\mathbb{C}P_\alpha^6$, which depends on the invariant of the elliptic curve.

EXERCISE. Verify similar properties of the deformations X_9 and J_{10} (with E_6 replaced by E_7 and E_8, respectively).

THEOREM (Looijenga). *The discriminant $\Delta_\tau \subset \mathbb{C}P_\alpha^\mu$ of a complex crystallographic group of type E_μ is holomorphically equivalent to the projectivized (on a hyperplane $a = $ const) discriminant $\nabla_{a(\tau)}$ of the corresponding parabolic singularity.*

The proof of this theorem is based on the period map, to which we shall devote the rest of this section.

REMARK. In singularity theory for two-dimensional complete intersections in $\mathbb{C}^4$ (see §4.2), the singularity Q_7 has a seven-parameter miniversal deformation with quasihomogeneous degrees of parameters (0 , 1 , 1 , 1 , 1 , 2 , 2). Its discriminant admits the same description in terms of the complex crystallographic groups of type D_5.

4.5. Monodromy [9, 4]. Let F_λ be a miniversal deformation of a parabolic singularity in three variables, $V_\lambda = F_\lambda^{-1}(0)$, $\lambda \in \mathbb{C}^\mu \setminus \Delta$ a nonsingular zero-level variety.

THEOREM (Gabrielov). *The intersection form $\langle \, , \, \rangle$ in an appropriate basis of the lattice $\mathbb{Z}^\mu = H_2(V_\lambda)$, $\mu = 8, 9, 10$, is the direct sum $0 \oplus E_{\mu-2}$ of the zero form in one variable and a negative semidefinite form corresponding to the extended Dynkin diagram of type $E_{\mu-2}$.*

Recall that the vertices of the diagram correspond to independent cycles with intersection number -2, cycles not connected by an edge are orthogonal, and those connected by an edge have intersection number 1. The form described by the Dynkin diagram has a one-dimensional kernel generated by a linear combination $\sum d_k v_k$ of basis vectors whose coefficients are just the weights depicted in the figure in §1.6. This form is negative definite on the quotient space with respect to the kernel, and isomorphic to the intersection form of a one-dimensional simple singularity, that is, it is defined by an ordinary (nonextended) Dynkin diagram. Therefore, the kernel of the intersection form of a parabolic singularity is two-dimensional.

Let us describe the action of the monodromy group of a parabolic singularity in the complex cohomology space $\mathbb{C}^\mu = H^2(V_\lambda, \mathbb{C})$ of a nonsingular level.

(1) The space $\mathbb{C}^\mu$ contains an integral invariant subspace $\mathbb{C}^{\mu-2}$, namely, the orthogonal complement of the kernel of the intersection form in the homology, on which the monodromy group acts as the corresponding Weyl group $E_{\mu-2}$.

(2) The monodromy group acts trivially on the quotient space $\mathbb{C}^2 = \mathbb{C}^\mu/\mathbb{C}^{\mu-2}$. This is because the monodromy group is generated by reflections relative to the intersection form of the cycles in hyperplanes orthogonal to the vectors of a certain integral basis (not the one used in the theorem). Therefore, its action in the homology is trivial on the kernel of the intersection

form, and in the cohomology it is trivial on the cokernel $\mathbb{C}^2$. In addition, the cokernel has an integral structure, namely, the lattice $\mathbb{Z}^2 \subset \mathbb{C}^2$, which is dual to the kernel in $H_2(V_\lambda, \mathbb{Z})$.

(3) Let (z_1, z_2) be coordinates in an integral basis of the cokernel $\mathbb{C}^2$. The inverse image of a point $z \in \mathbb{C}^2$ in $\mathbb{C}^\mu$ is an invariant affine subspace parallel to $\mathbb{C}^{\mu-2}$. The monodromy group acts on this space as a semidirect product G_z of the Weyl group $E_{\mu-2}$ and the translation group:

$$0 \to (z_1 \mathbb{Z}^{\mu-2} + z_2 \mathbb{Z}^{\mu-2}) \to G_z \to W \to 1.$$

(4) If $\tau = z_2/z_1$ is a rational number, then G_z is isomorphic to an affine Weyl group (see [1]), i.e., to the extension $0 \to \mathbb{Z}^{\mu-2} \to G \to W \to 1$ of W by the translation lattice which appears in the definition of the Weyl group.

(5) If τ is real, but irrational, then the lattice $z_1 \mathbb{Z}^{\mu-2} + z_2 \mathbb{Z}^{\mu-2}$ is dense in $z_1 \mathbb{R}^{\mu-2}$ and the group G_z is not discrete.

(6) If $\tau \notin \mathbb{R}$, then G_z is isomorphic to a complex crystallographic group of type $E_{\mu-2}$ and acts in the affine space $\mathbb{C}^{\mu-2}$ as $\tilde{W}$ with the parameter $\pm\tau$.

4.6. Invariant cycles. The cycles in $H_2(V_\lambda)$ that are invariant with respect to the monodromy group can be described by an explicit geometrical construction. Recall that a surface V_a with a parabolic singular point is obtained from a line bundle $L \to E_\tau$ ($a = a(\tau)$) over an elliptic curve by the compactification at infinity and removal of the zero section. Hence the neighborhood of infinity on $V_a \subset \mathbb{C}^3$ looks like a punctured neighborhood of the zero section in the bundle $(L \setminus E_\tau) \to E_\tau$. By choosing a suitable circle in each fiber (for example, the unit circle in the Hermitian metric) we obtain an S^1-bundle over E_τ. Let γ_1, γ_2 be a basis in $H_1(E_\tau)$, i.e., a meridian and a parallel on the torus E_τ. The union of the fibers of the S^1-bundle over γ_1, γ_2 are toroidal cycles Γ_1, Γ_2, which are independent on $V_a \setminus$ (singular point).

In a quasihomogeneous miniversal family, the functions

$$F_\lambda, \qquad \lambda = (a, \lambda_1, \ldots, \lambda_{\mu-1}),$$

with $a = a(\tau)$ fixed, have the same quasihomogeneous leading terms. Thus, the surfaces $V_\lambda = F_\lambda^{-1}(0)$ have the same structure in the neighborhood of infinity as the singular fiber V_a. Therefore, the toroidal cycles Γ_1, Γ_2 are well-defined on all surfaces V_λ and hence remain homologous to themselves under monodromy.

4.7. Structure of the period map [27]. Let $F_\lambda(x, y, z)$ be a quasi-homogeneous miniversal deformation of a parabolic point. Let us examine the period map of the 2-form $\omega = dx \wedge dy \wedge dz/dF_\lambda$,

$$\lambda \in (\mathbb{C}^\mu \setminus \Delta) \mapsto [\omega] \in H^2(V_\lambda, \mathbb{C}).$$

(1) Consider a one-parameter family of surfaces $V_{\lambda(t)}$ corresponding to one quasihomogeneous orbit

$$t \mapsto \lambda(t) = (a, t^{\alpha_1}\lambda_1, \ldots, t^{\alpha_{\mu-1}}\lambda_{\mu-1}).$$

If the surfaces $V_{\lambda(t)}$ are identified via quasihomogeneous dilatations, the classes $[\omega] \in H^2(V_{\lambda(t)}, \mathbb{C})$ are also identified. Indeed, ω is a quasihomogeneous form of degree 0 : $\frac{1}{3} + \frac{1}{3} + \frac{1}{3} = \frac{1}{2} + \frac{1}{4} + \frac{1}{4} = \frac{1}{2} + \frac{1}{3} + \frac{1}{6} = 1$. This means that on each hypersurface $\mathbb{C}^{\mu+1} = \{a = \text{const}\}$ the period map admits a quasihomogeneous projectivization:

$$\hat{\lambda} \in (\mathbb{C}P_\alpha^{\mu-2} \setminus \nabla_a) \mapsto [\omega] \in H^2(V_\lambda, \mathbb{C}).$$

(2) The image of this map lies in the hypersurface $z_2/z_1 = \tau$. Indeed, the integrals $z_i = \int_{\Gamma_i} \omega$, $i = 1, 2$, over invariant cycles can be computed in the neighborhood of infinity on V_2. This neighborhood is isomorphic to a punctured neighborhood of the zero section in the bundle $L \to E_\tau$, and the form ω in a first approximation (that is, on $V_{a(\tau)}$) is invariant with respect to dilatations in the fibers. Hence ω has a pole of first order on the zero section. Integrating ω over circles S^1 in the fibers of $L \to E_\tau$ is equivalent to computing the residue $2\pi i \operatorname{Res} \omega$ on the zero section. $\operatorname{Res} \omega$ is a holomorphic 1-form on the elliptic curve E_τ. It is therefore proportional to the 1-form dw on $\mathbb{C}/\mathbb{Z} \oplus \tau\mathbb{Z}$. Thus

$$z_i = \int_{\Gamma_i} \omega = 2\pi i \int_{\gamma_i} \operatorname{Res} \omega \sim \int_{\gamma_i} dw = 1 \text{ or } \tau,$$

so that the quotient of the invariant integrals is equal to the quotient of the periods of the elliptic curve.

(3) Associate with a class $[\omega]$ the straight line that it generates in $H^2(V_\lambda, \mathbb{C})$ and transfer this line by the Gauss-Manin connection to the cohomology of the distinguished surface V_{λ_0}. We obtain a holomorphic map

$$\mathbb{C}P_\alpha^{\mu-2} \setminus \nabla_{a(\tau)} \to \mathbb{C}_\tau^{\mu-2}/\tilde{W}$$

into the orbit space of a complex crystallographic group of type $E_{\mu-2}$.

THEOREM [27] *This map extends holomorphically to the discriminant and defines an isomorphism*

$$(\mathbb{C}P_\alpha^{\mu-2}, \nabla_{a(\tau)}) \to (\mathbb{C}_\tau^{\mu-2}/\tilde{W}, \Delta_\tau).$$

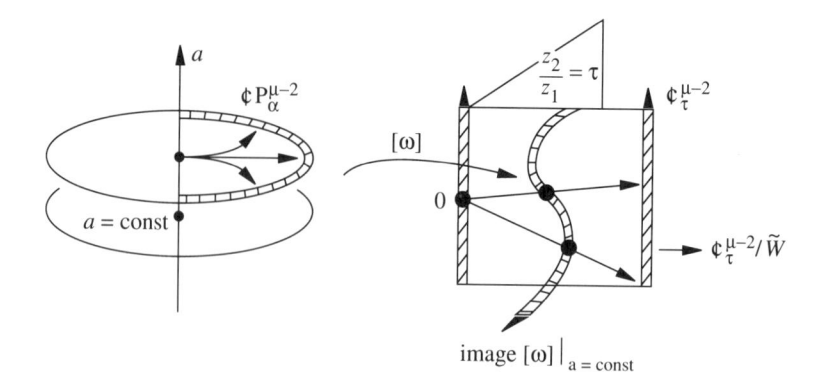

§5. Exceptional singularities

5.1. Fuchsian groups [20, 6, 11, 12, 10]. Consider a triangle on the Lobachevsky plane with angles $\left(\frac{\pi}{p}, \frac{\pi}{q}, \frac{\pi}{r}\right)$, where p, q, r are integers and $\frac{1}{p} + \frac{1}{q} + \frac{1}{r} < 1$. The reflections in the sides of the triangle generate a discrete group of motions. In the Poincaré model, the Lobachevsky plane is identified with the unit disk U in the complex plane. The holomorphic transformations form a subgroup of index 2 in our discrete group, namely, the Fuchsian group $\Phi \subset \operatorname{Aut} U = \operatorname{PSU}(1,1) \simeq \operatorname{PSL}_2(\mathbb{R})$. The fundamental domain of Φ consists of two mirror copies of the original triangle, and the quotient space U/Φ is a Riemann sphere with three "ramification points" of orders p, q, r, the vertices of the triangle.

The group Φ acts also on the cotangent bundle $L: T^*U \to U$. We define the surface $\hat{L}_\Phi$ as a one-point compactification of $(T^*U)/\Phi$, with the zero section U/Φ removed. The surface $\hat{L}_\Phi$ has one singular point at ∞.

DOLGACHEV'S THEOREM. *There exist exactly fourteen different triangles in the Lobachevskian plane such that the surface $\hat{L}_\Phi$ can be imbedded in $\mathbb{C}^3$. The equation $f(x,y,z) = 0$ of the image of the imbedding is quasihomogeneous and has an exceptional unimodal singularity at the singular point at ∞. All fourteen quasihomogeneous exceptional unimodal singularities are obtained in this way.*

The numbers (p, q, r) are called the *Dolgachev numbers* of the exceptional singularity. They are listed below.

K_{12} 2 3 7	Z_{13} 3 3 5	Q_{12} 3 3 6
K_{13} 2 4 4	W_{12} 2 2 5	S_{11} 2 5 6
K_{14} 3 3 4	W_{13} 3 4 4	S_{12} 3 4 5
Z_{11} 2 3 8	Q_{10} 2 3 9	U_{12} 4 4 4
Z_{12} 2 4 6	Q_{11} 2 4 7	

EXAMPLE. K_{12} (2, 3, 7). This example was studied in detail by Klein [6]. The Fuchsian group Φ contains a normal subgroup Φ_0 of index 168, which acts on U without fixed points. The map $U \to U/\Phi_0$ is a uniformization of the compact Riemann surface $C = U/\Phi_0$ of genus 3 with automorphism group Φ/Φ_0. The image of the canonical imbedding (i.e., the imbedding by holomorphic 1-forms)

$$C \hookrightarrow \mathbb{C}P^2 = P^*(\Omega^1 C)$$

is defined by the homogeneous equation $G = x_0^3 x_1 + x_1^3 x_2 + x_2^3 x_0 = 0$. The group Φ/Φ_0, a simple group of order 168 isomorphic to $\mathrm{PSL}_2(\mathbb{Z}_7) \simeq \mathrm{GL}_3(\mathbb{Z}_2)$, acts in $\mathbb{C}^3 = (\Omega^1 C)^*$ as a group of unitary reflections and preserves the cone $G^{-1}(0)$. The quotient space $\mathbb{C}^3/(\Phi/\Phi_0)$ is again isomorphic to $\mathbb{C}^3$ (see [19]). The basic invariants x, y, z have degrees (21, 14, 6), and the image of the cone has the equation $x^2 + y^3 + z^7 = 0$.

In the general case, the surface is equipped with an algebraic structure by defining the algebra of regular functions

$$\mathscr{A}_\Phi = \bigoplus_{q=0}^{\infty} \mathscr{A}_q(\Phi),$$

where $\mathscr{A}_q(\Phi)$ is the space of Φ-invariant holomorphic sections of the qth power L^q of the cotangent bundle $T^* U$. Such sections are called Φ-*automorphic forms of weight* q, and the entire space $\mathscr{A}_q(\Phi)$ is called the algebra of automorphic forms of Φ.

EXERCISES. (1) Prove that Φ-automorphic forms can be considered as meromorphic sections of the powers $T^{*q} C$ of the cotangent bundle of the Riemann surface $C = U/\Phi$ with poles only at ramification points.

(2) Let $p \in C$ be a ramification point of order ν. Prove that the section $T^{*q} C$, corresponding to an automorphic form of weight q, has a pole at p of order at most $n_\nu(q) = [q - q/\nu]$ (where [ ] is the integral part).

(3) Assume that $C = U/\Phi$ is compact, $(p_1, \ldots, p_n)$ are ramification points of orders $(\nu_1, \ldots, \nu_n)$. Show that $\mathscr{A}_q(\Phi)$ is isomorphic to the space of all meromorphic sections of $T^{*q} C$ with poles of order $\leq n_{\nu_i}(q)$ at the ramification points.

(4) Using the Riemann-Roch Theorem, show that if $q \geq 2$, then

$$\dim \mathscr{A}_q(\Phi) = (2q - 1)(g - 1) + \sum_{i=1}^{n} [q - q/\nu_i],$$

where $g = \dim \mathscr{A}_1(\Phi)$ is the genus of C.

(5) Let S be the area of the fundamental polygon of Φ in the Poincaré metric. Show that

$$\lim_{q \to \infty} \frac{\dim \mathscr{A}_q(\Phi)}{q} = \frac{S}{2\pi} = 2g - 2 + \sum_{i=1}^{n} \left(1 - \frac{1}{\nu_i}\right).$$

(6) Prove that the triangular group $(2, 3, 7)$ investigated by Klein has the minimal possible area of a fundamental domain among all Fuchsian groups with compact quotient space.

REMARK. It can also be shown (see [10]) that the group of maximal order among the automorphism groups of hyperbolic compact Riemann surfaces is the group of order 168 described in the above example.

5.2. **Intersection form** [20, 9, 5]. Let $T_{P,Q,R}$ denote a graph of the form

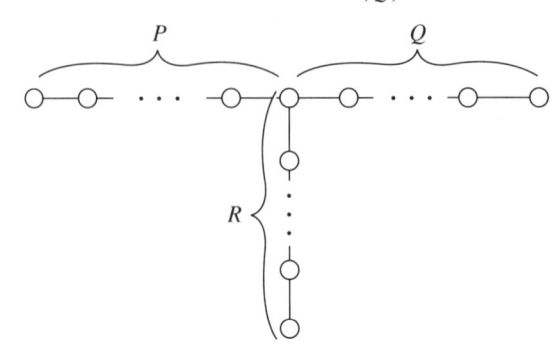

Define a bilinear symmetric form on a lattice $\mathbb{Z}^\mu$ of rank $P+Q+R-2$ by the usual rule $\langle \sigma, \sigma \rangle = -2$, $\langle \sigma_i, \sigma_j \rangle = 0$ or 1.

EXERCISE. Prove that the form $T_{P,Q,R}$ is negative definite if $\frac{1}{P} + \frac{1}{Q} + \frac{1}{R} > 1$ (Dynkin diagrams D_μ $(2, 2, \mu - 2)$ and E_μ, $(P, Q, R) = (2, 3, 3)$, $(2, 3, 4)$, or $(2, 3, 5)$); it is negative semidefinite if $\frac{1}{P} + \frac{1}{Q} + \frac{1}{R} = 1$ (extended Dynkin diagrams E_μ, $(P, Q, R) = (3, 3, 3)$, $(2, 4, 4)$, or $(2, 3, 6)$); it has a Minkowski signature if $\frac{1}{P} + \frac{1}{Q} + \frac{1}{R} < 1$.

THEOREM (Gabrielov). *The intersection form in the homology space $H_2(V_\lambda)$ of the nonsingular level surface of a versal deformation of an exceptional unimodal singularity is isomorphic to the direct sum $T_{P,Q,R} \oplus \begin{pmatrix} 0 & 1 \\ 1 & 0 \end{pmatrix}$ of an indefinite form of type T with a hyperbolic cell.*

COROLLARY. *The intersection form of an exceptional singularity is nondegenerate and its positive inertia index is* 2.

The numbers (P, Q, R) are called the *Gabrielov numbers* of the exceptional singularity. Their sum is the Milnor number $\mu = \dim H_2(V_\lambda)$. There exists a strange duality of exceptional singularities, noticed by D. B. Fuchs: Gabrielov numbers of one singularity are equal to Dolgachev numbers of another. The sum of the Milnor numbers of the singularities which are dual in this sense is 24 ($= p+q+r+P+Q+R$). The singularities K_{12}, W_{12}, Z_{12}, U_{12}, S_{12}, Q_{12} are self-dual, the others are divided into pairs of duals:

$$Q_{10} \leftrightarrow K_{14}, \quad Q_{11} \leftrightarrow Z_{13}, \quad S_{11} \leftrightarrow W_{13}, \quad Z_{11} \leftrightarrow K_{13}.$$

At the end of this section we shall present an explanation of this strange duality, due to Dolgachev–Nikulin and to Pinkham, which is based on the theory of $K3$-surfaces.

5.3. Cartan domains [13].
From Cartan's classification of bounded symmetric domains in complex spaces, we shall need the description of the so-called irreducible domains of type IV.

Let $\langle\,,\,\rangle$ be a symmetric bilinear form in $\mathbb{R}^{n+2}$ of signature $(+,+,-,\ldots,-)$. Consider the following subset in the complexification $\mathbb{C}^{n+2}$ of $\mathbb{R}^{n+2}$:

$$\{z \in \mathbb{C}^{n+2} | \langle z, z \rangle = 0, \ \langle z, \overline{z} \rangle > 0\}.$$

The projectivization of this set is an open subset of a quadric in $\mathbb{C}P^{n+1}$; it can be shown to be the union of two connected components, which differ in the sign of Im z_1/z_2. Let U^n be one of the components and Aut U^n the subgroup of index 2 in the automorphism group $O(2,n)$ of the form $\langle\,,\,\rangle$ that maps U^n into itself.

EXAMPLES. The domain U^1 is analytically isomorphic to the Poincaré disk, U^2 to the product of two Poincaré disks; when $n > 2$, U^n is irreducible.

The domain U^n is isomorphic to a bounded contractible subset of $\mathbb{C}^n$, and the group Aut U^n acts on it transitively and symmetrically. ([7])

EXERCISE. (1) Verify that the stabilizer of a point in U^n is a compact subgroup of Aut U^n.

(2) Let $\Gamma \subset$ Aut U^n be a discrete subgroup. Then Γ acts on U^n discretely (that is, Γ-orbits in U^n are discrete), and the quotient space possesses the structure of a complex variety, possibly noncompact and with singularities.

Let Γ be the monodromy group of an exceptional unimodal singularity that acts in the cohomology space $H^2(V^\lambda, \mathbb{C})$ of a nonsingular fiber of a miniversal deformation. The action of Γ preserves the form $\langle\,,\,\rangle$, dual to the intersection number of the cycles. As can be shown ([8]), Γ is a subgroup in Aut $U^{\mu-2}$ of the Cartan domain corresponding to the form $\langle\,,\,\rangle$. In addition, the monodromy group consists of integral transformations and consequently Γ is discrete in Aut $U^{\mu-2}$. Let B_Γ denote the orbit space $U^{\mu-2}/\Gamma$, and $\Delta_\Gamma \subset B_\Gamma$ the hypersurface of nonregular orbits, namely, orbits of fixed points of the transformations from Γ.

5.4. Structure of the discriminant [28, 33, 23].
A miniversal deformation of a singularity K_{12} has the form

$$x^3 + y^7 + z^2 + axy^5 + \lambda_1 xy^4 + \cdots + \lambda_{11}.$$

Among the monomials $(1, y, y^2, x, y^3, xy, y^4, xy^2, y^5, xy^3, xy^4, xy^5)$, which form a basis of the local algebra of the quasihomogeneous germ $f = x^3 + y^7 + z^2$, the last one has quasihomogeneous degree higher than that of

---

([7]) That is, it contains an involution with a single fixed point.

([8]) Verify that a reflection in a hyperplane that is orthogonal to a vector with negative "scalar square" belongs to Aut $U^{\mu-2}$.

f. Putting $a = 0$, we obtain an 11-parameter deformation of f, which we will call a *quasihomogeneous deformation*.

Miniversal deformations of the other exceptional singularities have the same structure; they contain a $(\mu - 1)$-parameter quasihomogeneous deformation F_λ, $\lambda = (\lambda_1, \ldots, \lambda_{\mu-1})$.

Let $\mathbb{C}P_\alpha^{\mu-2}$ be a quasihomogeneous projectivization of the parameter space of this deformation. Let B_S denote the open subset in $\mathbb{C}P_\alpha^{\mu-2}$ whose points correspond to surfaces $V_\lambda = F_\lambda^{-1}(0)$ with *only simple singularities* A, D, E; let $\Delta_S \subset B_S$ denote the discriminant, at whose points V_λ indeed has such singularities.

THEOREM 2. *There exists an analytic diffeomorphism $B_S \to B_\Gamma$ of the projectivized parameter space of a quasihomogeneous deformation of an exceptional unimodal singularity onto the orbit space of the monodromy group in its Cartan domain, which maps the discriminant Δ_S onto the variety Δ_Γ of nonregular orbits.*

COROLLARY. *The orbit space $B_\Gamma = U^{\mu-2}/\Gamma$ admits a compactification that completes it to a quasihomogeneous projective space $\mathbb{C}P_\alpha^{\mu-2}$.*

5.5. Structure of the period map. As usual, the form $\omega = dx \wedge dy \wedge dz/dF_\lambda$ defines a period map

$$\lambda \in (\mathbb{C}^\mu \setminus \Delta) \mapsto [\omega] \in H^2(V_\lambda, \mathbb{C})$$

which is a section of the cohomology bundle over the complement of the discriminant of a miniversal deformation of an exceptional singularity. It turns out that the restriction of $[\omega]$ to a quasihomogeneous deformation satisfies the conditions

$$\langle [\omega], [\omega] \rangle = 0, \quad \langle [\omega], [\overline{\omega}] \rangle > 0,$$

and therefore defines a map

$$(B_S \setminus \Delta_S) \to (U^{\mu-2}/\Gamma).$$

It is this map that extends to the isomorphism referred to in the previous theorem.

All surfaces $V_\lambda = f_\lambda^{-1}(0)$ of a quasihomogeneous family have the same structure at infinity. One can simultaneously compactify them at infinity and then resolve the singularities that arise there. It turns out that after this compactification ω can be extended to the resulting nonsingular compact surfaces $\hat{V}_\lambda$ as a holomorphic 2-form. Its integrals over the cycles that arise during compactification and resolution vanish. Hence $\langle [\omega], [\omega] \rangle$ and $\langle [\omega], [\overline{\omega}] \rangle$ can be computed as integrals over $\hat{V}_\lambda$ of the type

$$\int \omega \wedge \omega \quad \text{and} \quad \int \omega \wedge \overline{\omega},$$

thus explaining the above relationships.

5.6. Compactification and resolution at infinity.

The structure of V_0 (and hence of all surfaces V_λ in a quasihomogeneous deformation) at infinity is the same as that of the punctured neighborhood of the zero section in the quotient space T^*U/Φ of the cotangent bundle of the Poincaré disk modulo a triangular Fuchsian group of type (p, q, r). Our compactification amounts simply to restoring the zero section back to its place.

The compactified surfaces have three singular points at infinity, namely, ramification points on the Riemann sphere U/Φ. Let us ascertain the nature of these singularities. The stationary subgroup in Φ of a vertex of a fundamental triangle with angle π/ν is a finite cyclic group of order ν. In a neighborhood of the vertex on T^*U it acts as a cyclic subgroup of SU_2 on the plane $\mathbb{C}^2$. According to the results of §1.5, the quotient surface has a simple singularity of type $A_{\nu-1}$.

The resolution of simple singularities of series A was investigated in §3: the exceptional divisor consists of a set of projective lines that intersect in accordance with the Dynkin diagram. In addition to these three divisors, $\hat{V}_0$ contains the compactified projective line U/Φ (zero section), which intersects each of the three exceptional divisors once, at the corresponding ramification point. Thus, we have

THEOREM. *The divisor $\hat{V}_0 \setminus V_0$ on $\hat{V}_0$ is described by the graph $T_{p,q,r}$, where (p, q, r) are the Dolgachev numbers.*

COROLLARY. *The intersection form of 2-cycles on a nonsingular compact surface $\hat{V}_\lambda$ is isomorphic to the direct sum $T_{p,q,r} \oplus T_{P,Q,R} \oplus \begin{pmatrix} 0 & 1 \\ 1 & 0 \end{pmatrix}$, where (P, Q, R) are the Gabrielov numbers.*

Indeed, $H_2(\hat{V}_\lambda, \mathbb{R})$ is generated by algebraic cycles of the divisor $\hat{V}_\lambda \setminus V_\lambda$ and by finite cycles in $H_2(V_\lambda)$, which do not intersect them (verify!).

In particular, $\dim H_2(\hat{V}_\lambda) = p + q + r + P + Q + R - 2 = 22$ in view of the "strange duality".

5.7. $K3$-surfaces [14, 15].

A nonsingular simply connected closed complex surface is called a *surface of type $K3$* if there exists a nonvanishing holomorphic 2-form on it.

From a topological point of view, all $K3$-surfaces are identical. As a model we can take the quartic in $\mathbb{C}P^3$ whose homogeneous equation is $x_0^4 + x_1^4 + x_2^4 + x_3^4 = 0$. Its homology group is $H_2(V) \simeq \mathbb{Z}^{22}$; the intersection form has signature $(+, +, +, -, \ldots, -)$ and is isomorphic to the direct sum $3\begin{pmatrix} 0 & 1 \\ 1 & 0 \end{pmatrix} \oplus 2E_8$.

From an analytical point of view, $K3$-surfaces have moduli. All $K3$-surfaces are Kählerian, so that $H^2(V, \mathbb{C})$ has a Hodge decomposition $H^{2,0} \oplus H^{1,1} \oplus H^{0,2}$ by types of differential forms. The spaces $H^{2,0}$ and $H^{0,2}$ are one-dimensional and are generated by the nondegenerate holomorphic 2-form ω and the form $\bar{\omega}$, respectively. The space $H^{1,1}$ is the 20-dimensional orthogonal complement of $H^{2,0} \oplus H^{0,2}$ relative to the *Hermitian* form $\langle x, \bar{y}\rangle$ (the dual of the intersection form of the cycles). The restriction of $\langle x, \bar{x}\rangle$ to $H^{1,1}$ has signature $(+, -, \ldots, -)$.

The holomorphy type of a $K3$-surface is uniquely determined by the line $H^{2,0}$ in the space $H^2(V, \mathbb{C})$ of the complexified lattice $(H^2(V, \mathbb{Z}), \langle , \rangle)$. This line satisfies the following obvious restrictions:

$$\langle \omega, \omega\rangle = \int_V \omega \wedge \omega = 0, \qquad \langle \omega, \bar{\omega}\rangle = \int_V \omega \wedge \bar{\omega} > 0,$$

but is otherwise arbitrary. The domain B^{20} determined by these conditions on a quadric in the projective space is somewhat similar to a Cartan domain. The moduli space of $K3$-surfaces is therefore the quotient space of B^{20} by the integral automorphism group of the form $\langle , \rangle$. However, because of the "wrong" signature of the form $\langle , \rangle$, this discrete subgroup of $O(3, 19)$ acts on B^{20} in a nondiscrete way, and the moduli space is not Hausdorff.

Not all $K3$-surfaces are algebraic. The point is that on a surface V in $\mathbb{C}P^n$ a hyperplane cuts out a holomorphic curve C. For any holomorphic curve, obviously, $\int_C \omega = \int_C \bar{\omega} = 0$. Therefore C determines a nonzero integral cohomology class $[c]^* \in H^{1,1}$ (the Poincaré dual of its fundamental cycle). At the same time, if $H^{2,0}$ is generic in B^{20}, a subspace in $H^2(V, \mathbb{C})$ that is orthogonal to $H^{2,0} \oplus H^{0,2}$ does not contain nonzero integer points.

EXERCISE. Prove that $\langle [c]^*, [c]^*\rangle > 0$.

Algebraic $K3$-surfaces form a countable number of 19-parameter families. Any such family is determined by the choice of the positive integral class $[c]^*$ and in turn determines a hyperplane $B_c = \{\langle \omega, [c]^*\rangle = 0\}$ in B^{20}. This domain B_c is a 19-dimensional Cartan domain of type IV, and the moduli space of the algebraic $K3$-surfaces in a given family is obtained from B_c by factorization with respect to the automorphism subgroup of the lattice $(\mathbb{Z}^{22}, \langle , \rangle)$ that preserves $[c]$.

Moreover, the intersection of $H^{1,1}$ with the integral lattice is generated by the fundamental cycles of holomorphic curves on an algebraic $K3$-surface. A fixed intersection $D = H^{1,1} \cap H^2(V, \mathbb{Z}) \simeq \mathbb{Z}^k$ determines a $(20-k)$-dimensional Cartan domain

$$B_D = \{H^{2,0} \in B | \langle H^{2,0}, D\rangle = 0\}$$

and a $(20-k)$-parameter family of algebraic $K3$-surfaces with a common *divisor lattice* D. The moduli space of such surfaces is the quotient space of B_D by the stationary subgroup of D in the integral automorphism group of the intersection form on $H^2(V, \mathbb{Z})$.

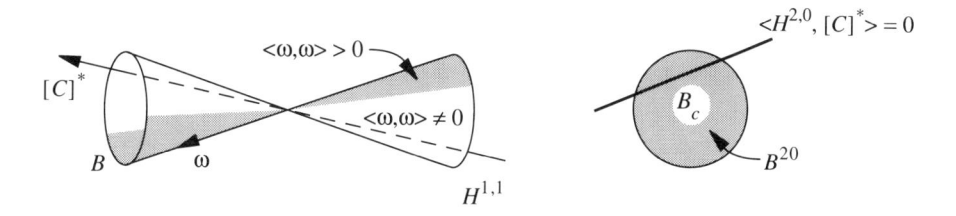

5.8. Strange duality [16, 34]. Returning to fourteen exceptional unimodal singularities, recall that we associated with each of them a family $\hat{V}_\lambda$ of nonsingular compact surfaces parametrized by points of the domain

$$B_S \setminus \Delta_S \subset \mathbb{C}P_\alpha^{\mu-2}.$$

THEOREM. *$\hat{V}_\lambda$ is a surface of type $K3$. The form $dx \wedge dy \wedge dz/dF_\lambda$ extends to $\hat{V}_\lambda$ as a holomorphic 2-form $\omega \in H^{2,0}$. The components of the divisor $\hat{V}_\lambda \setminus V_\lambda$ generate a lattice $D \subset H^{1,1} \cap H^2(\hat{V}_\lambda, \mathbb{Z})$ of type $T_{p,q,r}$ and dimension $p + q + r - 2 = 22 - \mu$. The family $\hat{V}_\lambda$ is a complete family of $K3$-surfaces with this divisor lattice.*

In addition, we obtain a decomposition ([9])

$$H^2(\hat{V}_\lambda) \cong 3 \begin{pmatrix} 0 & 1 \\ 1 & 0 \end{pmatrix} \oplus 2E_8 \approx \begin{pmatrix} 0 & 1 \\ 1 & 0 \end{pmatrix} \oplus T_{P,Q,R} \oplus T_{p,q,r}.$$

Of course, a generic cycle in $H^2(V_\lambda) = \begin{pmatrix} 0 & 1 \\ 1 & 0 \end{pmatrix} \oplus T_{P,Q,R}$ cannot be represented by a holomorphic curve in $\hat{V}_\lambda$. However, the very existence of the above decomposition shows that the "Gabrielov lattice" $T_{P,Q,R}$ is interchangeable with the "Dolgachev lattice" $T_{p,q,r}$, that is, it may be considered as the divisor lattice $D^* = H^{1,1} \cap \mathbb{Z}^{22}$. This choice determines a new Cartan domain $B_{D^*} \subset B^{20}$ (which in fact does not intersect the domain $B_D \subset B^{20}$) and thereby determines an algebraic family of $K3$-surfaces with the divisor lattice D^*. This new family is a quasihomogeneous deformation of the dual exceptional singularity.

REFERENCES

1. N. Bourbaki, *Eléments de mathématique. Groupes et algèbres de Lie.* IV–VI, Hermann, Paris, 1968.
2. _____, *Eléments de mathématique. Groupes et algèbres de Lie.* VII,VIII, Hermann, Paris, 1975.
3. V. I. Arnol'd, *Normal forms of functions near degenerate critical points, Weyl groups A_k, D_k, E_k and Lagrangian singularities*, Funktsional. Anal. i Prilozhen. **6** (1972), no. 2, 3–25; English transl. in Functional Anal. Appl. **6** (1972), no. 2.
4. V. I. Arnol'd, A. N. Varchenko, and S. M. Guseĭn-Zade, *Singularities of differentiable maps*, vol. I, "Nauka", Moscow, 1982; English transl., Birkhäuser, Basel, 1985.

([9]) The symbol $\approx$ denotes a quasimorphism: forms $T_{p,q,r}$ are in general not unimodular and the right-hand side of $\approx$ is included in the left-hand side as a subgroup of finite index.

5. _____, *Singularities of differentiable maps*, vol. II, "Nauka", Moscow, 1984; English transl., Birkhäuser, Basel, 1988.
6. F. Klein, *Vorlesungen über die Entwicklung der Mathematik in 19 Jahrhundert*, Springer-Verlag, Berlin, 1926.
7. J. Milnor, *Singular points of complex hypersurfaces*, Princeton Univ. Press, Princeton, N. J., 1968.
8. I. N. Bernstein and O. V. Shvartsman, *Chevalley theorem for complex crystallographic Coxeter groups*, Funktsional. Anal. i Prilozhen. **12** (1978), no. 4, 79–80; English transl. in Functional Anal. Appl. **12** (1978), no. 4.
9. A. M. Gabrielov, *Dynkin diagrams of unimodal singularities*, Funktsional. Anal. i Prilozhen. **8** (1974), no. 3, 1–6; English transl. in Functional Anal. Appl. **8** (1974), no. 3.
10. I. Kra, *Automorphic forms and Kleinian groups*, W. A. Benjamin, Reading, Mass., 1972.
11. I. V. Dolgachev, *Factor-conic singularities of complex spaces*, Funktsional. Anal. i Prilozhen. **8** (1974), no. 2, 75–76; English transl. in Functional Anal. Appl. **8** (1974), no. 2.
12. _____, *Automorphic forms and quasihomogeneous singularities*, Funktsional. Anal. i Prilozhen. **9** (1975), no. 2, 67–68; English transl. in Functional Anal. Appl. **9** (1975), no. 2.
13. I. I. Piatetskiĭ-Shapiro, *Geometry of classical domains and the theory of automorphic functions*, Fizmatgiz, Moscow, 1961; English transl., *Automorphic functions and the geometry of classical domains*, Math. Appl., vol. 8, Gordon and Breach, New York, 1969.
14. I. R. Shafarevich (ed.), *Algebraic surfaces*, Trudy Mat. Inst. Steklov. **75** (1965); English transl. in Proc. Steklov Inst. Math. **75** (1967).
15. I. I. Piatetskiĭ-Shapiro and I. R. Shafarevich, *Torelli theorem for algebraic surfaces of type* $K3$, Izv. Akad. Nauk SSSR **35** (1971), no. 3, 530–572; English transl. in Math. USSR-Izv. **5** (1971), no. 3.
16. I. V. Dolgachev and V. V. Nikulin, *Exceptional singularities of V. I. Arnol'd and $K3$-surfaces*, All-Union Topology Conference, Abstract of papers, Minsk, 1977. (Russian)
17. O. P. Shcherbak, *Wave fronts and groups of reflections*, Uspekhi Mat. Nauk **43** (1988), no. 3, 125–160; English transl. in Russian Math. Surveys **43** (1988).
18. A. B. Givental', *Singular Lagrangian manifolds and their Lagrangian maps*, Itogi nauki i techniki. Sovr. probl. mat. Noveĭshie Dostizheniya, vol. 33, VINITI, Moscow, 1988, pp. 55–112; English transl. in J. Soviet Math. **52** (1990).
19. T. A. Springer, *Invariant theory*, Springer-Verlag, Berlin, Heidelberg, and New York, 1977.
20. V. I. Arnol'd, *Critical points of smooth functions*, Proc. Internat. Congr. Math. (Vancouver, 1974), vol. 1, pp. 19–39.
21. P. Slodowy, *Simple singularities and simple algebraic groups*, Lecture Notes in Math., vol. 815, Springer-Verlag, Berlin and New York, 1980.
22. E. Brieskorn, *Singular elements of semisimple algebraic groups*, Proc. Internat. Congr. Math. (Nice, 1970), vol. 2, pp. 279–284.
23. _____, *The unfolding of exceptional singularities*, Nova Acta Leopoldina **52** (1981), no. 240, 65–93.
24. K. Saito, *Quasihomogene isolierte Singularitäten von Hyperflächen*, Invent. Math. **14** (1971), 123–142.
25. _____, *Einfach-elliptische Singularitäten*, Invent. Math. **23** (1974), 289–325.
26. E. Looijenga, *The complement of the bifurcation variety of a simple singularity*, Invent. Math. **32** (1974), 105–116.
27. _____, *On the semiuniversal deformation of a simple-elliptic hypersurface singularity. II*, Topology **17** (1977), 23–40.
28. _____, *Homogeneous spaces associated to unimodular singularities*, Proc. Internat. Congr. Math. (Helsinki, 1974), Acad. Sci. Fennica, Helsinki, 1980.
29. _____, *Root systems and elliptic curves*, Invent. Math. **38** (1976), 17–32.
30. M. Artin, *On isolated rational singularities of surfaces*, Amer. J. Math. **88** (1966), 129–136.
31. D. Bennequin, *Caustique mystique*, Seminaire Bourbaki, Exp. 634, Asterisque **133-134** (1986), 19–56.

32. H. C. Pinkham, *Simple-elliptic singularities, del Pezzo sufaces and Cremona transformations*, Several Complex Variables, Part 1, Proc. Symp. Pure. Math., vol. 30, Amer. Math. Soc., Providence, R. I., 1977, pp. 67–71.
33. _____, *Groupe de monodromie des singularités unimodulaires exceptionelles*, C. R. Acad. Sci. Paris Sér. I Math. **284** (1977), 1515–1518.
34. _____, *Singularités exceptionelles, la dualité étrange d'Arnold et les surfaces* $K3$, C. R. Acad. Sci. Paris Sér. I Math. **284** (1977), 615–618.

Translated by A. BOCHMAN

Qualitative Analysis of Singularly Perturbed Systems of Chemical Kinetics

V. M. GOL'DSHTEĬN AND V. A. SOBOLEV

Introduction

A wide variety of chemical processes is characterized by extreme differences in the rates of conversion of the reactants and in the rates at which the thermal characteristics and concentrations may vary. Here are some typical examples.

In catalytic systems, the rates of reactions taking place on the catalytic surface are one or more orders of magnitude higher than the reaction rates in the gas phase. The reactions on the catalyst surface usually take place at comparatively low temperatures, and heat is released at a rate substantially lower than the rate of changes in the reagent concentrations.

In combustion systems it is natural to find a high rate of heat release combined with a comparatively low rate of consumption of combustive material. In gas-phase systems this difference is so striking that the phenomenon of self-ignition of gas mixtures is known as "thermal explosion".

Moreover, it is a commonly recognized principle of chemical kinetics to classify the materials taking part in a process according to their conversion rates. Thus, one distinguishes between "active sites" and support substances.

Recent years have seen a growing interest in the influence of "buffer" phenomena on the main process. Examples of such phenomena in catalytic systems are aging of the catalyst and adsorption of nonreactive reaction products on the catalyst surface. Naturally, the rates of "buffer" phenomena are lower by an order of magnitude than the reaction rates.

Chemical systems that separate into "slow" and "fast" subsystems (according to their rates) are generally investigated by the method of quasi-steady-state concentrations. The idea is simple: one assumes that the system adjusts itself to the slow subsystem by quickly reaching equilibrium (quasi-steady state) in the fast subsystem. Thus a significantly smaller number of parameters is needed in the analysis of certain phenomena, and the model is considerably simplified. The question of the range of applicability of the Bodensteĭn-Semenov quasi-steady-state approximation remains open. (For a more detailed treatment of the method, see [1].)

One mathematical model of a chemical system separated into fast and slow subsystems is a singularly perturbed system of differential equations (i.e., a system in which the coefficient of some of the derivatives is a small parameter). We are interested in the case when the model is a system of ordinary equations of the form

$$\frac{dx}{dt} = f(x, y, t, \varepsilon), \qquad \varepsilon \frac{dy}{dt} = g(x, y, t, \varepsilon). \tag{1}$$

Here x and y are vectors and ε a small parameter ($0 \leq \varepsilon \ll 1$). It is assumed that the values of f and g are comparable with unity. The system of equations $\frac{dx}{dt} = f(x, y, t, \varepsilon)$ represents the slow subsystem, and the system of equations $\varepsilon \frac{dy}{dt} = g(x, y, t, \varepsilon)$ the fast subsystem; so it is natural to call $\frac{dx}{dt} = f(x, y, t, \varepsilon)$ the slow subsystem, and $\varepsilon \frac{dy}{dt} = g(x, y, t, \varepsilon)$ the fast subsystem of system (1).

In models of chemical kinetics the right-hand side of system (1) does not depend on time. In models of chemical reactors a time dependence may appear (e.g., in control of the process, allowance for external noise, and so on).

As applied to system (1), the quasi-steady-state approximation leads to the following procedure: set $\varepsilon = 0$ and study the system of equations $\frac{dx}{dt} = f(x, y, t, 0)$, on the assumption that x and y satisfy the equation $g(x, y, t, 0) = 0$.

Our interest (in the context of models of chemical processes) lies in qualitative analysis, which is concerned with steady-state phenomena, vibrational modes, domains of multiplicity of steady states, and so on. If the dimension of the system of equations (1) is greater than two, the methods of the qualitative theory of differential equations are generally very difficult to apply. This is especially evident in nonisothermic models, owing to the strong nonlinearity of the system. Numerical analysis is also of little avail when applied to such models when they involve multi-scaled variables (and this is precisely the situation in singularly perturbed systems), because of the "stiffness" of the system, that is, its high sensitivity to small calculation errors. On the other hand, the presence of the small parameter allows asymptotic methods to be used for its analysis. However, most asymptotic methods are intended for the solution of boundary- and initial-value problems over a finite time interval and are not suitable for qualitative analysis.

In the present paper we propose a method for qualitative asymptotic analysis of the kind of perturbed systems of differential equations that are relevant for chemical processes. The method relies on the theory of integral manifolds [2, 3], which essentially replaces the original system by another system, on an integral manifold, of dimension equal to that of the slow subsystem. In the zero-epsilon approximation ($\varepsilon = 0$) this method leads to a modification of the quasi-steady-state approximation.

To illustrate the concept of an integral manifold, consider the equation $g(x, y, t, 0) = 0$. It defines a surface in x, y, t space. In the quasi-steady-state approximation, one concentrates on the parts of this surface that can be defined by an explicit equation $y = \varphi(x, t)$ (that is, when the fast variables can be expressed in terms of the slow variables and time). If $g(x, y, t, 0)$ is nonlinear, the equation $g(x, y, t, 0) = 0$ cannot always be resolved. A sufficient condition is that the assumptions of the implicit function theorem be fulfilled ($\det(\partial g/\partial y)(x, y, t, 0) \neq 0$). The surface may split into several "sheets", each defined by an equation $y = \varphi(x, t)$. The method of quasi-steady-state concentrations tells us to insert $y = \varphi(x, t)$ into the slow subsystem and analyze the latter. This means that we are presupposing the existence of trajectories of system (1) that lie on the surface $g(x, y, t, 0) = 0$. Such trajectories, obviously, do not exist, since $\frac{dy}{dt} = 0$ at every point of the surface, and hence the trajectory has to "pierce" the surface.

Nevertheless, the quasi-steady-state approximation is applicable provided certain fairly lenient conditions are imposed on the function $g(x, y, t, \varepsilon)$, because at a distance of order $O(\varepsilon)$ from every sheet of the surface $g(x, y, t, 0) = 0$ there lies a surface (integral manifold) that contains trajectories of system (1). It is the existence of the integral manifold that justifies the quasi-steady-state approximation.

For an approximate qualitative analysis of system (1), we propose first to study the structure, the form, and singularities of the integral manifold by using asymptotic expansions in terms of ε. An algorithm for computing such expansions will be described.

The sheets of the (slow) surface $g(x, y, t, 0) = 0$ are taken as the zeroth approximation. A qualitative analysis of the slow subsystem is carried out in each sheet of the integral manifold. Depending on the degree of "crudeness" of the sheet and the qualitative picture on it, we can use the zeroth, first, or subsequent approximations of the integral manifold for our approximate analysis. For example, if a qualitative analysis of system (1) by quasi-steady-state concentrations reveals a steady state that is not crude (center or saddle point), then, in the first approximation, this state may either split into several different steady states or turn out to be crude, and so on.

It is now easy to understand the scheme of approximate qualitative analysis:

1) Analyze the form and singularities of each sheet of the integral manifold in the zeroth approximation.
2) Carry out a qualitative analysis of the slow subsystem on each sheet.
3) If the situation turns out to be "noncrude" in some sense, refine it by using the first approximation of the integral manifold.
4) If the situation turns out to be "noncrude" in the first approximation

as well, use further approximations.
5) Carry out a qualitative analysis of the fast subsystem.
6) Knowing the types and singularities of the sheets, using the qualitative picture on the sheets and the results of the qualitative analysis of the fast subsystem, establish an overall picture of the qualitative behavior of system (1).

At each stage except the last, both numerical and analytical tools can be used. The last stage does not presuppose the use of numerical analysis. The structure of the algorithm is such that numerical analysis is used only for subsystems with singly-scaled variables (the slow subsystems on the sheets of the integral manifold and the fast subsystem). This eliminates the main computational difficulties due to the "stiffness" of the original system. Analytical tools are used for subsystems of dimension considerably lower than that of the original system, and for these subsystems they turn out to be more efficient. For example, if the total dimension of system (1) is four and that of the slow subsystem two, qualitative analysis will be used only for two-dimensional subsystems.

This algorithm allows us to refine and modify A. M. Zhabotinskii's "scheme of a concentration system" (see [1]). In our opinion, Zhabotinskii's scheme, both in its original version and in the modification presented here, has a wider range of application than kinetic models of concentration systems. It is in fact a general scheme for applying approximate methods of qualitative analysis to the study of mathematical models with differently-scaled variables.

REMARK. Closely related to the method of integral manifolds is the subordination principle of synergetics, according to which some of the variables in a system can be expressed in terms of the others, provided that the spectrum of the linearized system satisfies certain assumptions.

Approximative qualitative analysis by the above method may concern itself with various types of dynamic behavior in system (1):

1) determination of domains of slow and fast motion;
2) relaxation trajectories, that is, trajectories in which a fast section follows a slow one;
3) steady states on stable sheets of the integral manifold;
4) oscillations of various types: relaxation; slow (limit cycles of the slow subsystem on stable sheets of the manifold); fast (limit cycles of the fast subsystem).

The list can be enlarged, if necessary, in each specific case.

§1. Elements of the geometric theory of singularly perturbed systems

We shall be concerned in this section mainly with systems of ordinary differential equations (1). As usual, x and y are vectors in the Euclidean spaces $\mathbb{R}^m$ and $\mathbb{R}^n$ (m and n are the dimensions of the spaces), $t \in (-\infty, +\infty)$ is a variable with the meaning of time, ε is a small positive parameter, $\dot{x}$

and $\dot{y}$ are derivatives with respect to time: $\dot{x} = dx/dt$, $\dot{y} = dy/dt$. In all cases of interest here, the vector-valued functions f, g are infinitely differentiable (at least, in the relevant domain). We shall, therefore, assume throughout that f and g are sufficiently smooth.

If system (1) can be reduced by a change of variables to the form

$$\dot{v} = F(v, t, \varepsilon), \qquad (2)$$

$$\varepsilon\dot{w} = G(v, w, t, \varepsilon), \qquad (3)$$

the slow subsystem can be studied independently. Qualitative analysis of the system as a whole will then split into several stages, each dealing with only one subsystem (fast or slow). Consequently, one of the main technical questions of this section is whether system (1) can indeed be reduced to the form (2), (3). We will describe an approximate method for calculating the necessary change of variables (by expansion in terms of the small parameter). This approach to the analysis of singularly perturbed systems is based on the theory of integral manifolds [2-4]. It has been successfully applied in the mechanics of systems of solid bodies and gyroscopes [3-4], and in the field of automatic control [5-6].

Our purpose is the qualitative investigation of system (1). As far as the study of nonrelaxation steady-state behavior and limit cycles is concerned, the problem reduces to analysis of the slow subsystem (2). In the analysis of relaxation oscillations one can use methods and results of [7].

Consider the "associated" subsystem, that is, $dy/d\tau = g(x, y, t, 0)$, $\tau = t/\varepsilon$, treating x, t as parameters. We shall assume that some of the steady states $y^0 = y^0(x, t)$ of this subsystem are asymptotically stable and that a trajectory starting at any point of the domain approaches one of these states as closely as desired as $t \to \infty$. This assumption will hold, for example, if the matrix $(\partial g/\partial y)(x, y^0(x, t), t, 0)$ is stable for part of the stationary states and the domain can be represented as the union of the domains of attraction of the asymptotically stable steady states.

The system

$$\dot{x} = f(x, y, t, 0), \qquad (4)$$

$$0 = g(x, y, t, 0) \qquad (5)$$

obtained from (1) by setting $\varepsilon = 0$, is called the *degenerate system*. Equation (5) defines an m-dimensional surface M, called the *slow surface*. The intersection of M with the surface $\det(\partial g/\partial y)(x, y, t, 0) = 0$ is an $(m-1)$-dimensional surface Γ which divides M into parts, in each of which $\det(\partial g/\partial y)(x, y, t, 0)$ has a fixed sign. It is natural to call each such part a sheet of the slow surface. The boundaries of the sheets are either parts of Γ or intersections of the slow surface with the boundary of the domain in which system (1) is being studied.

As a rule, a sheet is the graph of a well-defined vector-valued function $y = h^0(x, t)$. The slow surface may consist of several sheets, defined by different functions $y = h_i^{(0)}(x, t)$, whose domains may intersect, depending on the complexity of the structure of the slow surface. The number of sheets may be investigated by analyzing the singularities of the slow surface. If the dimension of the fast variables is one, the structure of the slow surface may be analyzed by applying elementary catastrophe theory to the function $g(x, y, 0)$ (provided the latter is independent of time). In that case the state space y is one-dimensional, and the space of control parameters (the variables x) has dimension m [8].

The boundaries of the sheets of the slow surface are defined by the equation $\det(\partial g/\partial y)(x, y, t, 0) = 0$. This relation means that the classification of the variables as fast or slow breaks down near the boundary of a sheet, so that the analysis of system (1) cannot be reduced to that of the fast or slow subsystem only. Since the conditions of the implicit function theorem do not hold on the boundary of a sheet, the asymptotic methods discussed below are useless there. In that case we confine ourselves to a hypothesis concerning the "crude" behavior of the system near the boundary of a sheet, treating the latter as the place where the trajectories undergo a transition from slow variation (at the "tempo" of the slow subsystem) to fast (at the "tempo" of the fast subsystem). Specifically, we shall proceed as if the trajectory "breaks away" from the boundary of the sheet [7]. The sheet boundary may therefore be divided into two parts: one in which the "breakaway" hypothesis applies and a "noncrude" part, in which the "breakaway" hypothesis fails to hold. The former will be called the breakaway curve or surface. The structure of the second part of the sheet boundary may be rather complicated. For example, it may be (part of) the self-intersection surface of the slow surface M. A detailed exposition of conditions under which the breakaway hypothesis is true can be found in [7].

For the degenerate system, we can introduce the concept of a discontinuous solution [7]. Formally speaking, the degenerate system has solutions on the slow surface only (more precisely, on a sheet of the slow surface). Even if the initial point of a trajectory lies on a sheet of the slow surface, the trajectory formally exists only up to the breakaway point. But if we consider a trajectory of the degenerate system to be the limit as $\varepsilon \to 0$ of trajectories of the original system, we arrive at the following interpretation of a discontinuous solution. If the initial point (x_0, y_0, t_0) is not on the slow surface, the solution changes instantaneously either leaving the domain of the fast subsystem or becoming a steady state of the fast subsystem. The change concerns only the fast variable. In the second case the solution jumps instantaneously from (x_0, y_0, t_0) to (x_0, y_1, t_0), where $g(x_0, y_1, t_0) = 0$. If (x_0, y_1, t_0) is not a breakaway point, the next part of the degenerate system changes on the sheet of the slow surface M that contains (x_0, y_1, t_0). The time t will then no longer be fixed, because the solution changes on the slow surface at the "tempo" of the slow subsystem, so the change cannot be instantaneous. If in the course of time a solution on a sheet reaches a breakaway point (x_1, y_1, t_1), it is again assumed that the solution changes instantaneously. Here, again, there are two alternatives: 1) the solution leaves the domain of the fast subsystem; 2) the solution becomes a steady state of the fast subsystem.

Other possible situations, when the solution reaches a point on the sheet boundary other than a breakaway point, will not be considered.

The transition from the original to the degenerate system naturally raises the question of the relation between the solutions of these systems.

If the phase trajectories of the original system are close to the corresponding phase trajectories of the degenerate system, over both fast and slow parts, then near every sheet of the slow surface there will exist a surface containing the slow parts of the trajectories of the original system. This assumption leads us to the concept of an integral manifold, through which a change of variables in the original system (1), which reduces it to the uncoupled form (2), (3), can be given a geometrical interpretation.

We briefly describe the method of integral manifolds as applied to system (1), $0 \le \varepsilon \le \varepsilon_0$. The constant ε_0 defines the range of the small parameter ε.

A smooth surface S in $\mathbb{R}^m \times \mathbb{R}^n \times \mathbb{R}$ is called an *integral manifold* of system (1) if any trajectory of the system that has at least one point in common with S lies entirely in S. Formally, if $(x(t_0), y(t_0), t_0) \in S$, then the trajectory $(x(t, \varepsilon), y(t, \varepsilon), t)$ lies entirely in S.

An integral manifold of an autonomous system

$$\dot{x} = f(x, y, \varepsilon), \qquad \varepsilon \dot{y} = g(x, y, \varepsilon) \qquad (6)$$

has the form $S_1 \times (-\infty, \infty)$, where S_1 is a surface in the phase space $\mathbb{R}^m \times \mathbb{R}^n$. For autonomous systems, therefore, it is natural to consider $S_1 \subset \mathbb{R}^m \times \mathbb{R}^n$ as the integral manifold. In such cases, instead of "integral manifold" one often uses the term "invariant manifold", which is more natural since S_1 does not change position in the phase space $\mathbb{R}^m \times \mathbb{R}^n \times \mathbb{R}$ under the action of system (6). In other words, the surface remains invariant as system (6) evolves in time.

The simplest example of an integral manifold is an integral curve of a system. The phase trajectories of a system, including its steady state cycles, are examples of invariant manifolds. The extended phase space $\mathbb{R}^m \times \mathbb{R}^n \times \mathbb{R}$ is also obviously an integral manifold.

The most important integral manifolds are those of comparatively low dimensions which also have some additional property. The most commonly used property is stability, that is, the capacity to attract trajectories of the system that do not belong to the integral manifold.

The specific property used, however, will depend on the purpose of the investigation.

In real situations it is generally difficult to construct an integral manifold that possesses the required property in the whole extended phase space. For this reason the integral manifolds are constructed only locally, that is, in a certain domain of the phase space. This motivates the following definition.

A smooth surface $S \subset \mathbb{R}^m \times \mathbb{R}^n \times \mathbb{R}$ is called a *local integral manifold* of system (1.1), (1.2), if, whenever S contains a point $(x(t_0), y(t_0), t_0)$ of a trajectory $(x(t), y(t), t)$, it contains a whole part of the trajectory, that is, there exists T, $t_0 < T \leq \infty$, such that $(x(t), y(t), t) \in S$, $t \in [t_0, T]$. In what follows, however, we shall continue to use the unmodified term "integral manifold".

The only integral manifolds of system (1.1), (1.2) of relevance here are those of dimension m (the dimension of the slow variables) that can be represented as the graphs of vector-valued functions $y = h(x, t, \varepsilon)$. We also stipulate that $h(x, t, 0) = h^{(0)}(x, t)$, where $h^{(0)}(x, t)$ is a function whose graph is a sheet of the slow surface, and we assume that $h(x, t, \varepsilon)$ is a sufficiently smooth function of ε.

In autonomous systems the integral manifolds will be graphs of functions $y = h(x, \varepsilon)$. In such cases m is the geometrical dimension of the manifold. Following a well-established tradition, we do not count the time t as an additional dimension in nonautonomous systems, so that in such systems an m-dimensional integral manifold is $(m + 1)$-dimensional in the geometrical sense. Moreover, an integral manifold may be viewed not as a fixed surface but as a parametric family of closely situated surfaces, the role of parameter being played by ε.

Integral manifolds of the above form are called manifolds of slow motions —the origin of this term lies in nonlinear mechanics. An integral manifold may be regarded as a surface where the phase velocity has a local minimum, that is, a surface characterized by the most persistent phase changes (motions). Integral manifolds of slow motions constitute a refinement of the sheets of the slow surface, obtained by taking the small parameter ε into consideration.

The motion along an integral manifold is governed by the equation

$$\dot{x} = f(x, h(x, t, \varepsilon), t, \varepsilon). \tag{7}$$

If $x(t, \varepsilon)$ is a solution of this equation, then the pair $x(t, \varepsilon)$, $y(t, \varepsilon)$, where $y(t, \varepsilon) = h(x(t, \varepsilon), t, \varepsilon)$, is a solution of the original system (1.1), (1.2), since it defines a trajectory on the integral manifold. In synergetics, the existence of a representation $y = h(x, t, \varepsilon)$, that is, the possibility of expressing the fast variables in terms of the slow ones, is the content of the subordination principle [9].

The relation

$$y = h(x, t, \varepsilon) \tag{8}$$

valid on the integral manifold, implies that

$$\varepsilon \frac{\partial h}{\partial t} + \varepsilon \frac{\partial h}{\partial x} f(x, h, t, \varepsilon) = g(x, h, t, \varepsilon). \tag{9}$$

This equation is obtained by substituting h for y in (1).

Setting $\varepsilon = 0$ in (7), (8), and (9), we obtain $\dot{x} = f(x, h(x, t, 0), t, 0)$, $y = h(x, t, 0)$, and $g(x, h(x, t, 0), t, 0) = 0$, respectively. Thus, in the zero-epsilon approximation, $y = h(x, t, \varepsilon)$ defines a sheet of the slow surface, while equations (8), (9) define the degenerate system (4), (5).

Recall that the degenerate system comes into play in the quasi-steady-state approximation. When this approximation is used, the small parameter ε is set to be equal to zero in the left-hand side only, and the degenerate system is analyzed in the following form:

$$\dot{x} = f(x, y, t, \varepsilon), \qquad 0 = g(x, y, t, \varepsilon).$$

If the system of equations $y = h(x, t, \varepsilon)$ defines an integral manifold of system (1), then a change of variables will "straighten out" the integral manifold, to obtain the m-dimensional hyperplane $w = 0$.

As mentioned in the introduction, in a neighborhood of an integral manifold the system of equations can be reduced to the form (2), (3), (5). The reduction is accomplished in two steps. First we "straighten out" the integral manifold, obtaining an m-dimensional hyperplane, by the change of variables $y = w + h(x, t, \varepsilon)$.

At the second step, another change of variables $x = v + \varepsilon H(v, w, t, \varepsilon)$ is carried out, to reduce the slow subsystem to the form $\dot{v} = F(v, t, \varepsilon)$, where $F(v, t, \varepsilon) = f(v, h(v, t, \varepsilon), t, \varepsilon)$ [5].

Solutions of the original system are related to solutions of system (2), (3) as follows:

$$\begin{aligned} x &= v + \varepsilon H(v, w, t, \varepsilon), \\ y &= w + h(x, t, \varepsilon) = w + h(v + \varepsilon H(v, w, t, \varepsilon), t, \varepsilon). \end{aligned} \qquad (10)$$

Since the right-hand side of subsystem (2) does not depend on w, subsystems (2) and (3) can be solved independently. We may thus view (10) as a change of variables, which decomposes (uncouples) the original system of equations in a neighborhood of the integral manifold of slow motions. The original system can be brought to the form (2), (3) because its solutions are superpositions of solutions of the fast subsystem and those of the slow subsystem. Not only the original system but also the initial values can be uncoupled. The initial values of the original system, $x(t_0) = x^0$, $y(t_0) = y^0$, are reduced by (10) to $w(t_0) = w^0 = y^0 - h(x^0, t_0, \varepsilon)$, $x^0 = v^0 + \varepsilon H(v^0, w^0, t_0, \varepsilon)$, $v^0 = v(t_0)$.

The initial values of the slow subsystem (2) can be found approximately by asymptotic expansion of $H(v, w, t, \varepsilon)$ in powers of the small parameter.

When the method of integral manifolds is being used to solve a specific problem, a central question is the calculation of the function $h(x, t, \varepsilon)$ in terms of which manifold is described. Exact calculation is generally impossible, and various approximations are necessary. One possibility is asymptotic expansion of $h(x, t, \varepsilon)$ in powers of the small parameter:

$$h(x, t, \varepsilon) = h^{(0)}(x, t) + \varepsilon h^{(1)}(x, t) + \cdots + \varepsilon^k h^{(k)}(x, t) + \cdots. \quad (11)$$

Inserting the formal series (11) into (9), we get

$$\varepsilon \sum_{k \geq 0} \varepsilon^k \frac{\partial h^{(k)}}{\partial t} + \varepsilon \left(\sum_{k \geq 0} \varepsilon^k \frac{\partial h^{(k)}}{\partial x} \right) f\left(x, \sum_{k \geq 0} \varepsilon^k h^{(k)}, t, \varepsilon \right) \\
= g\left(x, \sum_{k \geq 0} \varepsilon^k h^{(k)}, t, \varepsilon \right). \quad (12)$$

Expand f and g formally in asymptotic series:

$$f\left(x, \sum_{k \geq 0} \varepsilon^k h^{(k)}, t, \varepsilon \right) = \sum_{k \geq 0} \varepsilon^k f^{(k)}(x, h^{(0)}, \ldots, h^{(k)}, t),$$

$$g\left(x, \sum_{k \geq 0} \varepsilon^k h^{(k)}, t, \varepsilon \right) = B(x, t) \sum_{k \geq 1} \varepsilon^k h^{(k)} + \sum_{k \geq 1} \varepsilon^k g^{(k)}(x, h^{(0)}, \ldots, h^{(k-1)}, t).$$

In the asymptotic expansion of g we have used the fact that $g(x, h^{(0)}, t, 0) = 0$. The notation $B(x, t)$ stands for $(\partial g/\partial y)(x, h_{(0)}, t, 0)$. Inserting the formal expansions into (4.2) and successively equating the coefficients of like powers of ε, we get a chain of equalities, from which the coefficients $h^{(0)}, h^{(1)}, \ldots, h^{(k)}, \ldots$ of the formal series (11) can be successively determined. These equalities have the form

$$\frac{\partial h^{(k-1)}}{\partial t} + \sum_{0 \leq p \leq k-1} \frac{\partial h^{(p)}}{\partial x} f^{(k-1-p)} = B h^{(k)} + g^{(k)}.$$

Since B is a nonsingular matrix,

$$h^{(k)} = B^{-1} \left[\frac{\partial h^{(k-1)}}{\partial t} + \sum_{0 \leq p \leq k-1} \frac{\partial h^{(p)}}{\partial x} f^{(k-1-p)} - g^{(k)} \right].$$

The function H in (10) may also be represented as an asymptotic expansion in powers of ε [5].

As mentioned previously, attracting (stable) integral manifolds are of particular interest.

An integral manifold defined by a function $h(x, t, \varepsilon)$ such that the eigenvalues of the matrix $B(x, t) = (\partial g/\partial y)(x, h^{(0)}(x, t), t, 0)$ have negative real parts is stable. In other words, as t increases, any trajectory $(x(t, \varepsilon), y(t, \varepsilon), t)$ of the system that begins near the manifold approaches arbitrarily close to a trajectory on the manifold. More precisely, a solution $x = x(t, \varepsilon)$, $y = y(t, \varepsilon)$ of the original system satisfying the initial condition $x(t_0, \varepsilon) = x^0$, $y(t_0, \varepsilon) = y^0$ can be represented as

$$\begin{aligned} x(t, \varepsilon) &= v(t, \varepsilon) + \varepsilon \varphi_1(t, \varepsilon), \\ y(t, \varepsilon) &= \bar{y}(t, \varepsilon) + \varphi_2(t, \varepsilon). \end{aligned} \quad (13)$$

This representation is valid for initial data such that $|y^0 - h(x^0, y_0, \varepsilon)| < \rho$. If the integral manifold is stable, we can choose a constant ρ so small that there exists a point v^0 that is the initial data for a solution $v(t, \varepsilon)$ of the equation $\dot{v} = f(v, h(v, t, \varepsilon), t, \varepsilon)$; the functions $\varphi_1(t, \varepsilon)$, $\varphi_2(t, \varepsilon)$ are small corrections that determine the degree to which trajectories passing near the manifold tend asymptotically to the corresponding trajectories on the manifold as t increases. They satisfy the following inequalities:

$$|\varphi_i(t, \varepsilon)| \leq N|y^0 - h(x^0, t_0, \varepsilon)| \exp[-\beta(t - t_0)/\varepsilon], \qquad i = 1, 2, \quad (14)$$

for $t \geq t_0$. The positive constants N and β do not depend on the choice of x^0, y^0, t_0, and $\varepsilon \in (0, \varepsilon_0)$. The initial data v^0 depend on the choice of x^0, y^0, t_0, and can be approximated as an asymptotic expansion based on the formula

$$x^0 = v^0 + \varepsilon H(v^0, w^0, t_0, \varepsilon).$$

In view of the form of the change of variables (10) that uncouples the system, the exact expressions for φ_1 and φ_2 may be written as follows:

$$\varphi_1 = H(v(t, \varepsilon), w(t, \varepsilon), t, \varepsilon),$$
$$\varphi_2 = w(t, \varepsilon) + h(v(t, \varepsilon) + \varepsilon\varphi_1, t, \varepsilon) - h(v(t, \varepsilon), t, \varepsilon),$$

where $w(t, \varepsilon)$ is a solution of equation (3) with initial value $w(t_0, \varepsilon) = y^0 - h(x^0, t_0, \varepsilon)$.

From (13) and (14) we obtain the following reduction principle for a stable integral manifold defined by a function $y = h(x, t, \varepsilon)$: A solution $x = x(t, \varepsilon)$, $y = h(x(t, \varepsilon), t, \varepsilon)$ of the original system (1) is stable (asymptotically stable, unstable) if and only if the corresponding solution of the system of equations $\dot{v} = F(v, t, \varepsilon)$ on the integral manifold is stable (asymptotically stable, unstable).

The reduction principle and the representation (13) allow us to investigate the qualitative behavior of trajectories of the original system near the integral manifold by analyzing the equation on the manifold.

§2. The thermal explosion model

In this section we study the classical model of thermal explosion in the framework of a one-step, first-order reaction, with the consumption of the initial reactants allowed. By using the method of integral manifolds one can distinguish the main types of transient regimes and study the successive reorganization of reaction regimes when changes occur in the initial conditions.

In view of the special features of transient regimes, as distinct from slow and explosive regimes, the transitional region should be treated as a self-contained object of investigation, subject to laws of its own. Aspects of the evolution of reactions in the transitional region and other questions connected with critical phenomena in thermal explosion theory have been studied by several authors. Details may be found, for example, in [10].

By the self-ignition limit we mean the set of all initial conditions at which the reaction undergoes abrupt, jumplike transition from a slow to a faster (explosive) rate. In closed systems the "limit" is a narrow domain of parameters in which the nature of the reaction undergoes a radical transformation.

The system of differential equations, in the traditional dimensionless variables, is

$$\gamma\dot{\theta} = \eta\exp(\theta/(1+\beta\theta)) - \alpha\theta = g(\eta, \theta), \tag{15}$$

$$\dot{\eta} = -\eta\exp(\theta/(1+\beta\theta)) = f(\eta, \theta). \tag{16}$$

At high activation energies, with the high thermal effect of the reaction characteristic of explosive systems, the parameter γ takes fairly small values. The initial condition is $\eta(0) = 1$, $\theta(0) = 0$. The parameter α characterizes the initial state of the system and may vary over a broad range. The variable η should be interpreted as the dimensionless concentration of the reactant and θ as dimensionless temperature.

Dividing equation (15) by equation (16), we get the differential equation

$$\frac{d\theta}{d\eta} = -\frac{1}{\gamma} - \alpha\frac{\theta\exp(\theta/(1+\beta\theta))}{\eta}.$$

It is evident from this equation that $d\theta/d\eta \leq -1/\gamma$, that is, every trajectory of system (15)–(16) that intersects the adiabatic line $\eta = 1 - \gamma\theta$ in the phase plane (η, θ) remains below the line with increasing time. If $\theta = 0$, $\eta > 0$, then $0 < d\theta/dt$, and if $\eta = 0$, then $d\eta/dt = 0$. Hence, no trajectory of the system can leave the triangle U bounded by the coordinate axes $\eta = 0$, $\theta = 0$ and the adiabatic line $\eta = 1 - \gamma\theta$.

We shall study the system in the region of physical interest only, namely in the triangle U.

Since γ is small, it is natural to treat (15)–(16) as a singularly perturbed system. In the traditional technique, one lets $\gamma \to 0$ and investigates discontinuous solutions of the degenerate system. For standard singularly perturbed systems, these solutions are good approximations to the solutions of the original system, provided γ is small. However, in view of the considerable length of the triangle U (θ varies inside the triangle from 0 to $1/\gamma$), some care should be exercised in applying this method to the analysis of reactions with significant evolution of heat (that is, explosive regimes). The technique is nevertheless applicable in such cases as well, provided attention is confined to the domain of comparatively low evolution of heat.

The technique of discontinuous solutions defines the *self-ignition limit* as the specific initial data at which the reaction undergoes a *jumplike* transition from slow to explosive evolution. This approximate approach cannot recognize transient phenomena. That can be achieved by using the accurate approximations provided by asymptotic analysis.

Numerical calculations published by various authors confirm the existence of a very sharp transition in initial conditions from slow to explosive regimes.

Numerical investigation of the transitional region presents considerable difficulties. In the computational works known to the present authors, the smallest possible change in initial data that is tolerable within the limits of the method being used, results in a "jump" from slow to explosive evolution. In addition, the spread of the main reaction characteristics (maximal temperature, induction period, and so on) from slow to explosive regimes is enormous. We know of no published work rendering a comprehensive numerical account of the transitional region for typical parameter values—that is, a continuous transition from slow to explosive conditions; such a transition clearly exists, in view of the continuous dependence of the solution on the initial conditions.

The main difficulty in numerical work is high sensitivity of the solution to small variations of the initial conditions in the transitional region. This imposes stringent demands on the precision and stability of the numerical methods used; these demands become even more stringent the deeper one penetrates the transitional region.

To the best of our knowledge, the literature provides neither analytical studies of transients nor methods for their investigation. We propose to use one-dimensional integral manifolds for this purpose. Our major tool will be unstable integral manifolds.

The slow curve of system (15), (16) is defined by the condition $g(\eta, \theta) = 0$. This equation has a solution $\theta = \theta(\eta)$ everywhere, except at two breakaway points A and B. The curve thus has three branches, M_1, M_2, M_3, which are zeroth approximations of three one-dimensional integral manifolds S_1, S_2, S_3.

The manifolds S_1 and S_3 are stable, but S_2 is unstable. The branch S_1 connects the unique stable singular point of the system to the breakaway point A, and the branch S_2 connects the breakaway points A and B. The ordinate of A depends only on the parameter α.

Each of the integral manifolds S_1, S_2, S_3 is local, but it is also a part of some trajectory of the system. In what follows we shall use the same notation, S_1, S_2, S_3, for the entire trajectories.

If the parameter $\alpha = \alpha_0$ is suitably chosen, the trajectory containing S_2 will begin at $\eta_1 = 1$, $\theta = 0$ and describe a reaction that can naturally be considered critical, in the following sense: the reaction is not slow, because the evolution of temperature is substantially greater than unity; but neither is it explosive, since the temperature increases at the "tempo" of the motion along S_2, which is governed by the slow equation (15). There clearly exist trajectories that remain close to the critical trajectory over long time intervals. Depending on whether they lie above or below the critical trajectory, they will either "break away" in an explosion or move rapidly into a "slow" region. Analysis of the transient regimes requires the use of not only the zeroth approximation to S_2, but also higher-order approximations. Thus, transient regimes are represented by solutions whose trajectories lie along the unstable integral manifold S_2 over sufficiently large intervals. The crucial point is the identification of the critical trajectory as the one containing S_2. One might explain the "reorganization" of the solution as the transition from solutions on the stable integral manifold S_1, through S_2, to solutions close to the adiabatic line.

The curves S_2 and S_3 are also critical trajectories of the system: S_2, which we call the *slow critical trajectory*, separates slow reactions proper from transients. Trajectories between S_1 and S_2 correspond to slow transient regimes, characterized by comparatively fast (but not explosive) reactions, reaching fairly high conversion rates, followed by abrupt deceleration and subsequent transition to a slow reaction.

However, at slow rates the deceleration occurs at low conversion rates.

The initial state of the system, defined by the value α_1 of the parameter α at which S_1 passes through the point $\eta = 1$, $\theta = 0$, is also a self-ignition limit, which separates regions of slow transients.

The third critical trajectory S_3 does not intersect the η axis and therefore constitutes a new self-ignition limit. We only observe that a low trajectory, starting at the point $\eta = 1$, $\theta = 0$, does not pass below S_3.

The parameter value α_0 corresponding to the critical trajectory S_2, i.e., the self-ignition limit, is given by the following formula:

$$\alpha_0 = e(1-\beta)\left[1 - 2.946\left(1 + \frac{7}{3}\beta\right)\gamma^{2/3}\right] + \frac{4}{9}(1+6\beta)\gamma \ln\frac{1}{\gamma} + O(\gamma + \beta^2).$$

§3. Dynamical properties of catalytic reactors

Recent years have seen a growing interest in mathematical models of catalytic continuous stirred tank reactors in complex situations, due primarily to new experimental data concerning the complex dynamic behavior of chemical reactors (in particular, different types of self-oscillations). Self-oscillations are generally characterized by significant changes in the concentrations of reactants and reaction products. There has been a substantial gap between nonisothermic models with simple kinetics, which have been investigated in detail, with allowed changes in the concentrations of the reactants, and models of complex heterogeneous catalytic reactions, in which the concentrations of the reactants and the reaction products are assumed to be constant.

The purpose of this section is to give a detailed analysis of the dynamics of an isothermic continuous stirring tank reactor for a model reaction—oxidation of a material with simple molecules. The reaction is oxidation of CO on group VIII metals. The mechanism is assumed to correspond to a two-route scheme. One route is an adsorption mechanism, the other a shock mechanism. In the adsorption mechanism, two materials adsorbed on the surface of the catalyst interact. In the shock mechanism a material in the gaseous phase interacts with the molecules of another material adsorbed on the surface of the catalyst. Many authors believe there is a broad range of parameters in which the shock mechanism may not occur. In this case the general reaction mechanism would be adsorption.

In our model it is assumed that the total pressure of the reaction mixture remains constant throughout the reaction.

Our analysis of the dynamics of the process will use qualitative methods, based on investigating limiting behavior (asymptotics). We will be interested in the following main aspects of dynamics: steady states, types of steady states, conditions of multiplicity, relaxation regimes (abrupt changes of dynamical state), oscillations of various types, including relaxation oscillations. In our analysis of oscillatory behavior we shall not concern ourselves with the determination of limit cycles, but with a cruder situation: existence of solutions with constantly alternating fast and slow parts. The results of the present section were obtained jointly by the authors and L. I. Kononenko.

We will consider the detailed mechanism of a bimolecular reaction on the catalyst surface. There are two possible reaction mechanisms: a reaction between reactants adsorbed on the catalyst surface (adsorption mechanism); and a reaction between an adsorbed reactant and a reactant in the gaseous phase (shock mechanism). It is assumed that the reaction takes place at constant pressure.

The kinetic scheme of the reaction is as follows:

1) $(N_1)_2 + 2Z \underset{}{\overset{K_1}{\rightleftarrows}} 2N_1Z$,
2) $N_2 + Z \underset{}{\overset{K_2}{\rightleftarrows}} N_2Z$,
3) $N_1Z + N_2Z \overset{K_3}{\rightarrow}$ product (adsorption mechanism),
4) $N_2 + N_1Z \overset{K_4}{\rightarrow}$ product (shock mechanism).

This description is used for catalytic oxidation reactions, such as oxidation of CO on metals of the platinum group.

We use the following notation: $(N_1)_2$, N_2 are the reactants fed into the reactor; N_1Z, N_2Z the same reactants adsorbed on the catalyst surface; Z the free sites on the catalyst surface. Since the number of active sites on the catalyst surface is assumed to be constant, it follows that $[Z]+[N_1Z]+[N_2Z] = C$ (square brackets denote concentrations). The constant C is the number of active sites per unit area of the catalyst surface.

Let V_p be the volume of the gas in the reactor, S the area of the catalyst surface, V_0 the input velocity of the gaseous mixture, V the flow velocity at the reactor outlet; K_1, K_{-1}, K_2, K_{-2}, K_3, K_4 kinetic constants; A, B the initial concentrations of $(N_1)_2$, N_2. Let w_3 denote the reaction rate in stage 3, that is, $w_3 = K_3[N_1Z][N_2Z]$. At constant pressure, the flow velocity at the reactor outlet can be calculated by the formula

$$V = V_0 + (w_1 + w_2 + w_3)S. \qquad (17)$$

In dimensionless variables we have

$$x_1 = \frac{V_0}{V_p}[(N_1)_2], \quad x_2 = \frac{V_0}{V_p}[N_2]; \quad y_1 = [N_1 Z]/C, \quad y_2 = [N_2 Z]/C.$$

The system of equations in dimensionless variables must allow for the specific form (17) of the flow velocity at the reactor outlet as a function of the reaction rates.

We let $\omega_1 = \kappa_1 x_1 (1-y_1-y_2)^2 - \kappa_{-1} y_1^2$, $\omega_2 = \kappa_2 x_2 (1-y_1-y_2) - \kappa_{-2} y_2$, $\omega_3 = y_1 y_2$, $\omega_4 = \kappa_4 y_1 x_2$ denote the dimensionless reaction rates in stages 1–4. The parameters involved in the system are:

$$\kappa_1 = K_1/K_3, \quad \kappa_{-1} = K_{-1}/K_3, \quad \kappa_2 = K_2/K_3 C,$$
$$\kappa_{-2} = K_{-2}/K_3 C, \quad \kappa_4 = K_4/K_3 C,$$
$$\beta = K_3 C V_p/V_0, \quad \alpha = K_3 S C^2/V_p, \quad a = V_0 A/V_p, \quad b = V_0 B/V_p.$$

The range of variables is

$$W = \{(x_1, x_2, y_1, y_2) \mid 0 \le x_1 \le a, \ 0 \le x_2 \le b,$$
$$0 \le y_1, \ 0 \le y_2, \ y_1 + y_2 \le 1\}$$

The reaction rate on the catalyst surface is much higher than the adsorption rate. Hence, $K_3 \gg K_1, K_2$, that is, $\kappa_1, \kappa_2 \ll 1$. The desorption constants are assumed to be negligible in comparison with the adsorption constants. In addition, $\alpha \ll \beta$.

The parameters $\kappa_1, \kappa_2, \kappa_4, \kappa_{-1}, \kappa_{-2}$ satisfy the relations $\kappa_1, \kappa_2 \gg \kappa_4, \kappa_{-1}, \kappa_{-2}$; $\kappa_1 \ll 1$; $\kappa_2 \ll 1$. This means that the desorption stages are assumed to be much slower than the adsorption stages. In addition, it is assumed that interactions governed by the shock mechanism are comparatively slow. We are interested in situations in which $\kappa_4, \kappa_{-1}, \kappa_{-2} \gg \varepsilon = 1/\beta$. In view of these assumptions we can set

$$\kappa_{-1} = \mu \kappa_{-1}^0, \quad \kappa_{-2} = \mu \kappa_{-2}^0, \quad \kappa_4 = \mu \kappa_4^0,$$

where μ is a small parameter and $\kappa_{-1}^0, \kappa_{-2}^0, \kappa_4^0$ are quantities of order $O(1)$. In dimensionless variables the system is

$$\dot{x} = f(x, y, \nu, \mu), \quad \varepsilon \dot{y} = g(x, y, \nu, \mu), \tag{18}$$

where $x = (x_1, x_2)^T$, $y = (y_1, y_2)^T$, $\nu = (a, b, \kappa_1, \kappa_2, \kappa_{-1}^0, \kappa_{-2}^0, \kappa_4^0)^T$, $f = (f_1, f_2)^T$, $g = (g_1, g_2)^T$, and ε, μ are independent small parameters. The functions f_1, f_2, g_1, g_2 are of the form

$$f_1 = a - x_1 - \alpha(\omega_1 + (\omega_3 - \omega_1 - \omega_2)x_1),$$
$$f_2 = b - x_2 - \alpha(\omega_2 + \omega_4 + (\omega_3 - \omega_1 - \omega_2)x_2),$$
$$g_1 = 2\omega_1 - \omega_3 - \omega_4,$$
$$g_2 = \omega_2 - \omega_3.$$

The slow surface of system (18) is defined by the equations $g(x, y, \nu, \mu) = 0$, or in expanded form,

$$\alpha\omega_1 = \omega_3 - \omega_4, \qquad \omega_2 = \omega_3.$$

The slow surface consists of 10 sheets. But if $\kappa_{-1} \neq 0$, $\kappa_4 \neq 0$, only the sheets of the slow surface are in the domain W; if $\kappa_1 = \kappa_4 = 0$ there are four (in this case the sheet $y_1 = 0$, $y_2 = 0$ is on the boundary of W). In the zero-order approximation ($\mu = 0$), a sheet S_1 is defined by the equation $y_1 = 0$, $y_2 = 1$, and a sheet S_3 by $y_1 = (P + \sqrt{Q})/2$, $y_2 = (P - \sqrt{Q})/2$; $P = 1 - \kappa_2 x_2/2\kappa_1 x_1$, $Q = P^2 - 2\kappa_2^2 x_2^2/\kappa_1 x_1$. The sheets S_2 and S_4 are unstable.

The dimensionless concentrations x_1, x_2 of the reactants and the reaction rate $\omega_3 = y_1 y_2$ are observable magnitudes. Later, therefore, we shall try to interpret the results of our analysis in terms of these characteristics of the process.

In the crude approximation considered here, the dynamics of the system does not depend on the phase portrait on unstable sheets of the integral manifold. A knowledge of the positions of the unstable sheets is necessary to determine the domains of attraction of the stable sheets, but as we are dealing with qualitative analysis only and not with the domains of attraction of the stable sheets, we shall not touch on the structure of the unstable sheets. That is why we must reject the usual approach, which begins any analysis of the dynamics of a system by determining the number of steady states.

In the physically relevant domain W there exist only two stable sheets S_1, S_3 of the slow surface with $\kappa_{-1} \neq 0$, $\kappa_4 \neq 0$. On S_1 the quantity $\omega_3 = y_1 y_2$ is small. Therefore, in steady stationary states lying on this sheet the reaction proceeds at a slow rate. On sheet S_3, however, $\omega_3 = y_1 y_2$ is fairly large. Hence, in steady states on S_3 one observes a comparatively fast reaction rate. On each sheet one of the following situations may occur:

A) There are no steady states on the sheet.
B) There is a unique steady state on the sheet, which is a stable node.
C) There are two steady states on the sheet, one of which is a saddle point, the other a stable node.

Various combinations of these three situations on S_1 and S_3 determine the dynamics of the system. Let us analyze some of the situations that may arise.

1) There is a unique steady state on S_1, and no steady states on S_3. Depending on the initial values a and b of the concentrations x_1 and x_2, two situations are possible. In the first, after a short induction period, during which x_1 and x_2 remain almost constant, the reaction rate slowly stabilizes at a relatively low level. This is the situation when the initial data $(a, b, 0, 0)$ lie in the domain of attraction of S_1. In such cases the trajectory of the system rapidly reaches a neighborhood of S_1 and is then slowly attracted by the unique steady state on that sheet.

In the second situation, after a short induction period during which x_1 and x_2 remain constant, the reaction rate stabilizes at a fairly high level. Then both the reaction rate and the concentrations x_1 and x_2 begin to change slowly. After some time, the rate falls abruptly to a low level and then starts to stabilize slowly. This is the situation when the initial data lie in the domain of attraction of S_3. Under these conditions the trajectory of the system rapidly reaches a neighborhood of S_3 and then begins to change, moving slowly along the sheet. As there are no steady states on S_3, the trajectory ultimately breaks away from the sheet (this corresponds to an abrupt drop in the reaction rate). After this, it is attracted by the single stable steady state of the sheet.

Other modes of dynamical behavior will now be outlined in less detail.

2) There is a unique steady state on S_3. There are no steady states on S_1.

This situation is dual to that of case 1). The unique steady state on S_3 is stable. There are two possibilities. In the first case, after an induction period, the reaction rate slowly stabilizes, remaining at a fairly high level. In the second, the reaction rate first becomes low. It then changes slowly and abruptly increases, subsequently slowly stabilizing at a fairly high level.

As we see in cases 1) and 2), the existence of relaxation trajectories implies the existence of relaxation changes in the reaction rate (that is, a rapid change after comparatively stable behavior). We shall refer to this behavior of the reaction rate as relaxation.

3) There is a unique steady state on each of the sheets S_1, S_3. The steady state on each sheet is necessarily stable.

Depending on the initial data, the rate stabilizes either at a comparatively high level (in a domain of attraction of S_3) or at a low level (domain of attraction of S_1). Relaxation of reaction rate may occur. However, the problem has not been studied in detail.

4) There are a unique steady state on S_1 (stable node) and two steady states on S_3 (stable node and saddle point).

The behavior of the system is similar to that in case 3). The existence of a saddle point on S_3 implies that the reaction rate may undergo relaxation. In trajectories that pass near the saddle point, the changes in reaction rate may decelerate—in numerical computations this may produce the impression that the reaction rate is stabilizing.

5) There are a unique steady state (stable node) on S_3 and two steady states on S_1 (stable node and a saddle point).

This situation is nearly the same as in case 4), except that slow relaxation is due to the saddle point on S_1. Thus, on some trajectories the changes in reaction rate are decelerated. Unlike case 4), this occurs at a time when the reaction rate is not high.

The remaining modes of dynamical behavior can be analyzed in a similar way. We shall dwell in detail only upon the cases in which relaxation oscillations occur.

6) There are no steady states on S_1 and S_3. This is possible only if stage 2 is reversible, that is, if $\kappa_{-2} \neq 0$. Relaxation oscillations of the reaction rate ω_3 are associated with a relaxation limit cycle of the system. The reaction rate varies periodically by jumps, first from a high to a low value and then vice versa. The role of the feedback, necessary for the formation of oscillations in reaction rate, is played by the common effect of the nonlinearity of the system and the reversibility of adsorption at the second stage.

7) There are no steady states on S_1. On S_3 there are two steady states (stable node and saddle point). We will not go into a detailed description of the dynamics in this situation. It is sufficient to say that for certain positions of the steady states the system has a relaxation limit cycle, corresponding to which are relaxation oscillations of the reaction rate. If the saddle point is sufficiently close to the limit cycle trajectory, the reaction rate may decelerate, that is, relaxation oscillations of the reaction rate are combined with slow relaxation.

We have described, in a general way, the most interesting modes of dynamic behavior of the system. We point out some other features of the model. Our asymptotic analysis has shown that the desorption constant in the shock mechanism, κ_4, does not affect the qualitative behavior of the system, if $\kappa_1 \ll 1$ and $\kappa_4 \ll 1$.

Observe that on the slow surface the system has the form

$$\dot{x}_1 = a - x_1 - \alpha\omega_3(1 - x_1)/2, \qquad \dot{x}_2 = b - x_2 - \omega_3(2 - x_2)/2.$$

For the sheet S_1 we have $\alpha\omega_3/2 = \gamma_1 x_1/x_2^2$, and for the sheet S_3, $\alpha\omega_3/2 = \gamma_2 x_2^2/x_1$. Since $\gamma_1 = \alpha\kappa_2^2/4\kappa_1$ and $\gamma_2 = \alpha\kappa_{-2}^2\kappa_1/\kappa_2^2$, we obtain $\gamma_1\gamma_2 = \alpha^2\kappa_{-2}^2/4$. Thus, in order to measure the desorption rate constant κ_{-2}, we need only know how to measure γ_1 and γ_2. This can be done because $2\omega_1 = \omega_3$ on the slow surface, so we can express γ_1 and γ_2 in terms of observables: the reaction rate ω_3 and the concentrations x_1, x_2 of the reactants in the gaseous phase. To calculate γ_2, one measures the observed rate ω_3 and the concentrations $\bar{x}_1, \bar{x}_2$ at a low steady-state reaction rate (corresponding to the sheet S_1). Then

$$\gamma_2 = \alpha\bar{\omega}_3\bar{x}_2^2/2\bar{x}_1.$$

To compute γ_1, one measures the observed rate $\tilde{\omega}_3$ and concentrations $\tilde{x}_1, \tilde{x}_2$ at a high steady-state reaction rate (corresponding to the sheet S_3). Then

$$\gamma_1 = \alpha\tilde{\omega}_3\tilde{x}_2^2/2\tilde{x}_1.$$

Finally,

$$\kappa_{-2} = \frac{\bar{x}_2}{\tilde{x}_2}\sqrt{\frac{\bar{\omega}_1\tilde{\omega}_3\tilde{x}_1}{\bar{x}_1}}. \qquad (19)$$

Our conclusions may be summarized as follows.

1. Allowance for desorption at the first stage (oxygen desorption) and the shock mechanism, on the assumption that their kinetic constants are small, has no effect on the qualitative behavior of the model. The model may therefore be simplified by assuming that these stages are absent.

2. If desorption in the second stage (that is, desorption of CO) is ignored, no relaxation oscillations will occur in the model. The reversibility of the desorption κ_{-2} and the nonlinearity of the reaction rate probably constitute the feedback necessary for relaxation oscillation to occur.

3. The model yields a method for calculating the desorption rate constant κ_{-2} in terms of the reaction rates for two different steady-state regimes and the corresponding concentrations of the reactants in the gaseous phase (formula (19)).

4. The model simulates a multiplicity of stable steady states. Steady-state behavior is often observed after relaxation of the reaction rate (abrupt changes in rate after a long period of comparatively slow variation).

5. The model simulates relaxation oscillations. These relaxation oscillations present a rather complicated picture, involving a deceleration of the changes in reaction rate (corresponding to a relaxation limit cycle trajectory passing near a saddle point).

A detailed treatment of the above topics may be found in the authors' book [11].

References

1. A. M. Zhabotinskiĭ, *Concentration Self-Oscillations*, "Nauka", Moscow, 1974. (Russian)
2. Yu. A. Mitropol'skiĭ and O. B. Lykova, *Integral Manifolds in Nonlinear Mechanics*, "Nauka", Moscow, 1973. (Russian)
3. V. V. Strygin and V. A. Sobolev, *Separation of Motions by the Method of Integral Manifolds*, "Nauka", Moscow, 1988. (Russian)
4. V. A. Sobolev, *Fast and slow motions of gyroscopic systems*, Periodica Polytechnica, Electr. Engrg. **29** (1985), no. 1, Budapest, 57–66.
5. _____, *Integral manifolds and decomposition of singularly perturbed systems*, Systems Control Lett. **5** (1984), no. 3, 169–179.
6. P. V. Kokotovic, *Recent trends in feedback design: an overview*, Automatica **21** (1985), no. 3, 225–236.
7. E. F. Mishchenko and N. Kh. Rozov, *Differential Equations with Small Parameters and Relaxation Oscillations*, "Nauka", Moscow, 1975; English transl., Plenum, New York, 1980.
8. V. I. Arnol'd, *Contact structure, relaxation oscillations, and singular points of implicit differential equations*, Geometry and Theory of Singularities in Nonlinear Equations, Voronezh State University, Voronezh, 1987, pp. 3–8. (Russian)
9. H. Haken, *Advanced Synergetics: Instability Hierarchies of Self-Organizing Systems and Devices*, Springer-Verlag, Berlin, 1983.
10. V. I. Babushok and V. M. Gol'dsteĭn, *Self-ignition limit: transients of a reaction*, Preprint No. 10, Institute of Mathematics, Siberian Branch of USSR Academy of Sciences, Novosibirsk, 1985. (Russian)
11. V. M. Gol'dshteĭn and V. A. Sobolev, *Qualitative Analysis of Singularly Perturbed Systems*, Institute of Mathematics, Siberian Branch of USSR Academy of Sciences, Novosibirsk, 1988. (Russian)

Translated by L. G. HANIN

Bifurcations with Symmetries

V. V. GORYUNOV

In many fields of mathematics and physics one has to classify mappings that are symmetric with respect to the action of a certain group, and to study properties of such mappings. The present lecture considers some aspects of the theory of singularities of such mappings. We consider the classification and topological properties of functions that are invariant under an involution, under symmetric groups, or under arbitrary finite groups. Some general propositions about equivariant singularities (versatility theorems, finite determinacy theorems, and others) will be presented. We also look into symmetry breaking and formulate a number of assertions relating to phase transitions in physics.

By equivalence of invariant functions we shall always mean equivalence up to diffeomorphisms of the preimage that commute with the action of a given symmetry group.

§1. $\mathbb{Z}_2$-invariants

We begin with the simplest and best studied case in the theory of singularities. The classification of germs of functions that are invariant under reflections in a hyperplane and the analysis of the topology of their level sets are closely related to the Lie groups B_k, C_k, and F_4, whose Dynkin diagrams have double edges. This relationship is similar to that observed between the groups A_k, D_k, E_k and singularities of nonsymmetric functions on smooth manifolds (see the paper by A. B. Giventalʹ in this volume).

1. **Classification.** Recall that a germ of a function in a given function space is *simple* with respect to a given equivalence relation if sufficiently small perturbations of the germ give representatives of only finitely many equivalence classes.

Let σ be an involution acting on a germ of a smooth n-dimensional linear space with coordinates $x_1, \ldots, x_n$, in such a way that it changes the sign of the first coordinate. The invariants of σ are even functions of x_1.

THEOREM [1]. *The simple C^∞ (or real analytic) invariant functions on the germ $(\mathbb{R}^n, 0)$ are exhausted up to equivalence by the following list*:

$A_k, k \geq 0: \pm x_1^2 \pm x^{k+1} + q,$ $E_6: \pm x_1^2 + x_2^3 \pm x_3^4 + q,$

$B_k, k \geq 2: \pm x_1^{2k} \pm x_2^2 + q,$ $E_7: \pm x_1^2 + x_2^3 + x_2 x_3^3 + q,$

$C_k, k \geq 3: x_1^2 x_2 \pm x_2^k + q,$ $E_8: \pm x_1^2 + x_2^3 + x_3^5 + q,$

$D_k, k \geq 4: \pm x_1^2 + x_2^2 x_3 \pm x_3^{k-1} + q,$ $F_4: \pm x_1^4 + x_2^3 + q.$

Here $q = \pm x_r^2 \pm \cdots \pm x_n^2$, where $r = 4$ for singularities D_k, E_k, and $r = 3$ in the other cases.

In the holomorphic situation the symbol "$\pm$" in the formulation of the theorem should be replaced everywhere by "+".

REMARK. The set of nonsimple functions has codimension $n + 3$ in the space of σ-invariants such that $f(0) = 0$.

We say that a singularity of class X *adjoins* a singularity of class Y, $X \to Y$, if an arbitrarily small deformation of a germ in X yields a germ in Y.

An adjacency diagram for simple complex σ-invariants is as follows:

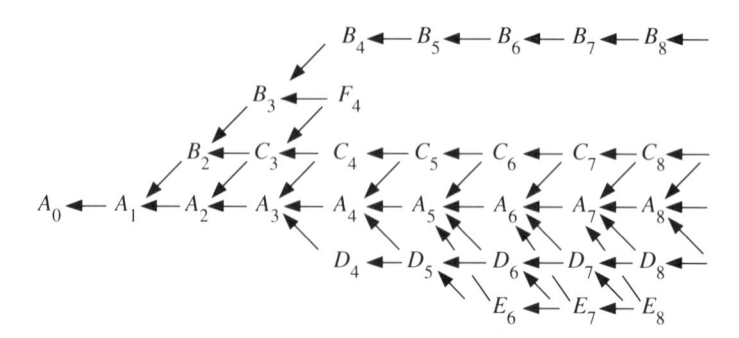

$$\begin{array}{c} B_4 \leftarrow B_5 \leftarrow B_6 \leftarrow B_7 \leftarrow B_8 \leftarrow \\ B_3 \leftarrow F_4 \\ B_2 \leftarrow C_3 \leftarrow C_4 \leftarrow C_5 \leftarrow C_6 \leftarrow C_7 \leftarrow C_8 \leftarrow \\ A_0 \leftarrow A_1 \leftarrow A_2 \leftarrow A_3 \leftarrow A_4 \leftarrow A_5 \leftarrow A_6 \leftarrow A_7 \leftarrow A_8 \leftarrow \\ D_4 \leftarrow D_5 \leftarrow D_6 \leftarrow D_7 \leftarrow D_8 \leftarrow \\ E_6 \leftarrow E_7 \leftarrow E_8 \end{array}$$

In the complex case, $\mathbb{C}^n$ is a twofold covering of the quotient space $\mathbb{C}^n/\sigma \simeq \mathbb{C}^n$. Take $y, x_2, \ldots, x_n$, where $y = x_1^2$, as coordinates in the quotient space. The plane $y = 0$, i.e., the image of the fixed points under factorization, is the ramification manifold of the covering. Any germ of an invariant function is of the form $f(x_1^2, x_2, \ldots, x_n)$. By factorization we associate it with the germ $f(y, x_2, \ldots, x_n)$. A biholomorphism of $\mathbb{C}^n$ that commutes with the involution corresponds under factorization to a biholomorphism of $\mathbb{C}^n/\sigma$ that maps the plane $y = 0$ into itself; and vice versa.

DEFINITION. A germ of the pair $(\mathbb{C}^n, \mathbb{C}^{n-1})$ is called a germ of a *complex manifold with smooth boundary*.

We see that the classification of the germs of σ-invariants is equivalent to the classification of function germs on a manifold with boundary with respect to the group of biholomorphisms of the manifold with boundary. Hence the above theorem also classifies the simple functions on a manifold with boundary (in particular, in the real case). This is how the subject is treated in [1]. On the whole, the theory of functions on a manifold with boundary is parallel to the theory of functions that are invariant under reflection in a hyperplane.

2. **Versal deformations.** We recall several general definitions.

A *deformation* of a point f in a space $\mathscr{F}$ is a germ of a smooth mapping F from a manifold Λ (called the base of the deformation) into $\mathscr{F}$ at a point 0 of Λ, such that $F(0) = f$.

Let $\mathscr{G}$ be a Lie group acting on $\mathscr{F}$. A deformation F of a point f is *versal* if any other deformation F' of the same point, with base Λ', can be represented as

$$F'(\lambda') = g(\lambda')F(\varphi(\lambda')), \qquad \lambda' \in \Lambda',$$

where g is a deformation of the identity element of $\mathscr{G}$, and $\varphi : (\Lambda', 0) \to (\Lambda, 0)$ is a smooth mapping (see Figure 1).

Thus, a versal deformation contains all possible deformations of the point under consideration.

A versal deformation is *miniuniversal* if its base has the smallest possible dimension.

For finite-dimensional manifolds $\mathscr{F}$, for the action of the diffeomorphism group of the preimage on the space of nonsymmetric function germs, and for a number of other "sufficiently good" problems in the theory of singularities (see, for example, §3 below), a miniversal deformation is a minimal transversal to the $\mathscr{G}$-orbit of the point. The same holds for σ-invariants.

Let $\mathscr{O}_n^\sigma$ be the space of germs at zero of C^∞ or real-analytic (or holomorphic) even functions of x_1. Let $F: (\mathbb{R}^n \times \mathbb{R}^\ell, 0) \to \mathbb{R}$ be a germ of an ℓ-parameter family of such functions, i.e., a deformation of a σ-invariant germ $f(x) = F(x, 0)$. Let $\lambda = (\lambda_1, \ldots, \lambda_\ell)$ be the deformation parameter. Denote the velocities of the initial deformation by $\dot{F}_i$.

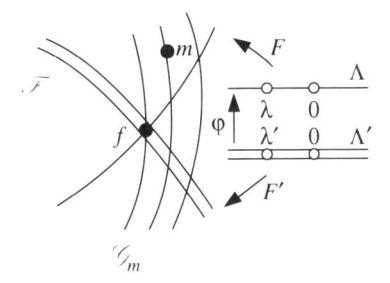

FIGURE 1

THEOREM. *F is a versal deformation in the class of σ-invariants if and only if*

$$\mathcal{O}_n^\sigma = \mathcal{O}_n^\sigma \langle x_1 \partial f/\partial x_1, \partial f/\partial x_2, \ldots, \partial f/\partial x_n \rangle + \mathbb{R}\langle \dot{F}_1, \ldots, \dot{F}_\ell \rangle,$$

where the angular brackets $\langle \cdots \rangle$ denote the module generated by the enclosed elements. The first module is the tangent space to the equivalence class of f.

Thus, a miniversal deformation of a σ-invariant f can be taken as $F(x, \lambda) = f(x) + \lambda_1 e_1(x) + \cdots + \lambda_\nu e_\nu(x)$, where $e_1, \ldots, e_\nu$ are representatives of a basis of the local ring

$$\mathcal{O}_n^\sigma / \mathcal{O}_n^\sigma \langle x_1 \partial f/\partial x_1, \partial f/\partial x_2, \ldots, \partial f/\partial x_n \rangle.$$

For simple functions, the subscript in the notation of the singularity is the codimension ν of the equivalence class.

3. Bifurcation diagrams of a σ-invariant.

DEFINITION. The germ at zero of the hyperspace $\Sigma \subset \Lambda$ of parameter values λ of a miniversal deformation F for which 0 is a critical value of the function $F(\cdot, \lambda)$ is called the *discriminant* (or *bifurcation diagram of zeroes*) of $f = F(\cdot, 0)$.

In the general case, the discriminant has two components: Σ_1 corresponds to the appearance of one critical point at the level $F(\cdot, \lambda) = 0$ on the plane of fixed points $x_1 = 0$, while Σ_2 corresponds to the appearance of a pair of symmetric critical points away from $x_1 = 0$. Figure 2 shows discriminants of real singularities for which the number of parameters of the miniversal deformation is at most three. The discriminants B_3 and $C_3^\pm$ are parts of the same complex surface—a cuspidal edge with a plane tangent to it.

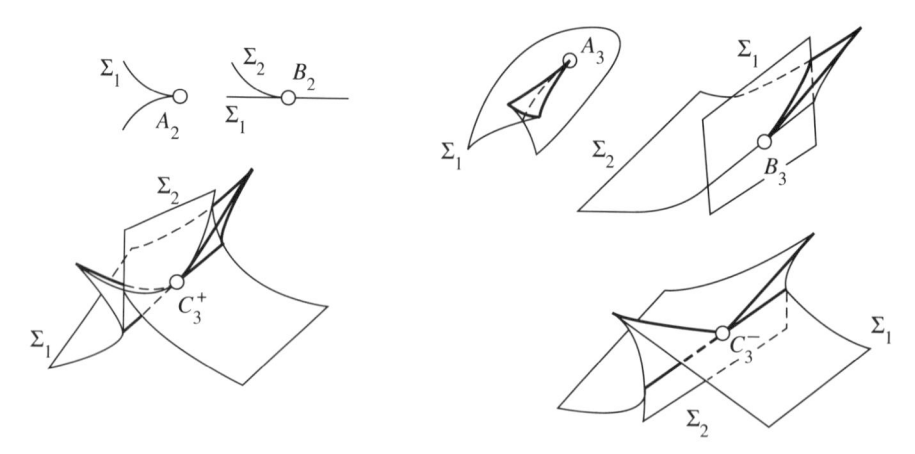

FIGURE 2

THEOREM [1]. *The discriminant of a simple complex σ-invariant singularity imbedded in C^ν is biholomorphically equivalent to the manifold of irregular orbits of the corresponding group, which is generated by reflections [3] acting on the complexification of Euclidean space.*

Thus, the complement of the discriminant of a simple singularity is the space $K(\pi, 1)$, where π is the Brieskorn braid group [2] determined by the corresponding Weyl group.

This result is a generalization to our case of a similar result for simple nonsymmetric functions.

The theorem about the homotopy type of the complement of the bifurcation diagram for functions with a simple singularity can also be extended from the nonsymmetric case to σ-invariants. Let us recall the definition of the bifurcation diagram.

It is easy to show that under a generic sufficiently small displacement of a germ of a complex σ-invariant function, a critical value decays into precisely ν different values.

DEFINITION. The *bifurcation diagram of functions of a complex σ-invariant* is the germ of the hypersurface $\Xi \subset \Lambda$ of the parameter values λ of a miniversal deformation F for which the function $F(\cdot, \lambda)$ has less than ν different critical values.

The diagram thus defined is cylindrical with respect to the deformation parameter corresponding to the element $e_\nu = 1$ of the local ring.

THEOREM [4]. *The complement $\Lambda\backslash\Xi$ of the bifurcation diagram of a simple complex σ-invariant function is the space $K(\pi, 1)$. The fundamental group π is a subgroup of index $\nu!h^\nu|W|^{-1}$ of the braid group on ν strings.*

Here h is the Coxeter number and $|W|$ the order of the corresponding Weyl group.

4. Nonsingular fiber.

Fix a representative $F(x, \lambda)$ of a symmetric versal deformation of the germ of an invariant function $f(x)$ at $0 \in \mathbb{C}^n$. Choose a small ball $B_\rho \subset \mathbb{C}^n$ of radius ρ centered at zero that contains no critical points of f other than 0. Choose δ so small that for $|\lambda| < \delta$, the sets $\{x : F(x, \lambda) = \delta\} \subset \mathbb{C}^n$ are transversal to the boundary of B_ρ. For λ in the ball $|\lambda| < \delta$, consider the *local level set* $V_\lambda = \{x : F(x, \lambda) = 0\} \cap B_\rho$. For nondiscriminant values of λ, these manifolds are nonsingular and diffeomorphic to each other. Each of them is homotopy equivalent to the union of μ $(n-1)$-dimensional spheres, where $\mu = \dim_\mathbb{C} \mathscr{O}_n/J(f)$ is the Milnor number of f ($\mathscr{O}_n$ is the space of germs of a holomorphic function on $\mathbb{C}^n$ at zero, and $J(f)$ is the ideal generated by partial derivatives $\partial f/\partial x_i$). Therefore the only nontrivial integral homology group of the nonsingular fiber V_λ is $H_{n-1} = \mathbb{Z}^\mu$. Let us construct a basis of this group.

Fix a nondiscriminant point λ_0, $\lambda_0 \notin \Xi$. Then $F(\cdot, \lambda_0)$ has exactly μ Morse critical points and ν different critical values. We may assume for the sake of convenience that the coordinates in the base of the deformation are chosen in such a way that to the one-parameter family of functions $\{F(\cdot, \lambda_0) + t\}$ with parameter t there corresponds a straight line $\ell \subset \Lambda$. This complex line intersects the discriminant at ν points corresponding to the critical values of $F(\cdot, \lambda_0)$. Let ν_1 (ν_2) be the number of these points in the intersection with the component Σ_1 (Σ_2). It is easy to show that

$$\nu_1 = \dim_{\mathbb{C}} \mathscr{O}_{n-1}/J(f') \quad \text{where } f' = f|_{x_1=0},$$

and

$$\nu_2 = \dim_{\mathbb{C}} \mathscr{O}_n^\sigma / \mathscr{O}_n^\sigma \langle x_1^{-1} \partial f/\partial x_1, \partial f/\partial x_2, \ldots, \partial f/\partial x_n \rangle = \dim_{\mathbb{C}} \mathscr{O}_n/J(f_*),$$

where in the latter case $\mathscr{O}_n$ are functions in the variables $y, x_2, \ldots, x_n$, and the germ $f_*(y, x_2, \ldots, x_n)$ is obtained from the representation $f(x_1, x_2, \ldots, x_n) = f_*(x_1^2, x_2, \ldots, x_n)$.

We have $\nu + 2\nu_2 = \mu$.

Let γ be a path on ℓ, going from λ_0 to one of the points, say λ_*, at which ℓ intersects the discriminant. With every point of this path we can associate a local level set that is singular only for the endpoint λ_*. If $\lambda_* \in \Sigma_1$ then, having reached λ_* along γ, we attach a symmetric $(n-1)$-dimensional sphere to the nonsingular fiber V_{λ_0}. The involution σ acts on this sphere as a reflection in the equator. If $\lambda_* \in \Sigma_2$, we attach a symmetric pair of spheres, and σ interchanges these spheres.

Now consider a collection of ν nonintersecting paths on ℓ from λ_0 to all points at which ℓ intersects Σ. This collection determines on the nonsingular fiber a set of vanishing spheres, which is the desired basis of the lattice $H_{n-1}(V_{\lambda_0}; \mathbb{Z})$.

5. **Anti-invariant homology.** The map σ acting on the nonsingular fiber V_{λ_0} defines an involution on the homology of the fiber. Let H be the part of H_{n-1} that is anti-invariant under this involution. It follows from our previous discussion of the action of σ on the vanishing spheres that $\dim_{\mathbb{Z}} H^- = \nu_1 + \nu_2$. For a basis in H_- we can take ν_1 spheres that vanish when we approach points of $\ell \cap \Sigma_1$ along the system of nonintersecting paths and ν_2 differences of symmetric pairs of spheres that vanish when we approach points of $\ell \cap \Sigma_2$ (the orientation is chosen appropriately). Such a basis is called a *distinguished* basis.

DEFINITION [1]. The homology class of a symmetric vanishing sphere in H^- is called a *short cycle*; an element of H^-, which is the difference of two symmetric spheres on V_{λ_0}, is called a *long cycle*.

The self-intersection numbers of short and long cycles for $n \equiv 3 \bmod 4$ are -2 and -4, respectively.

Consider a closed loop in $\Lambda \backslash \Sigma$. Each pass along this loop induces a map of the nonsingular level manifold onto itself, which commutes with σ. This yields the action on H^- of the monodromy group, i.e., of the image of

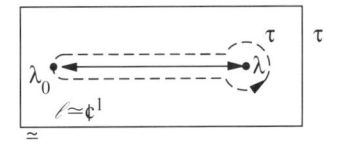

FIGURE 3

the natural representation of the fundamental group of the space $\Delta\backslash\Sigma$ as the group of automorphisms of H^- that preserve intersection numbers of cycles. The generators of the fundamental group can be obtained using the collection of nonintersecting paths, which was used to construct the distinguished basis of H^-. Figure 3 shows how to construct a loop for every path γ on ℓ from λ_0 to a point λ_* in $\ell \cap \Sigma$. Such a loop τ is called a *simple* loop.

Let e be an element of the distinguished basis of H^- corresponding to a path γ. The monodromy operator associated with the simple loop $\tau(\gamma)$ acts on H^- according to the generalized Picard-Lefschetz formulas:

$$a \mapsto a + (-1)^{n(n+1)/2}(a, e)e, \qquad \text{if } e \text{ is a short cycle;}$$
$$a \mapsto a + (-1)^{n(n+1)/2}(a, e)e/2, \qquad \text{if } e \text{ is a long cycle.}$$

Starting from the distinguished basis of H^-, we can construct the Dynkin diagram of a symmetric singularity, as follows. Let the dimension n be odd. To every vertex of the diagram there corresponds an element of the distinguished basis. Two vertices are connected by an edge of multiplicity k if the intersection number of the corresponding cycles is k and at least one of them is short, or if the intersection number of the two corresponding long cycles is $2k$. Edges connecting vertices corresponding to cycles of different length are labeled with the sign $>$, with the arrow pointing to the short cycle.

For simple σ-invariant singularities, there are distinguished bases in which the Dynkin diagrams of the singularities look like the canonical diagrams of the corresponding Weyl groups (Figure 4).

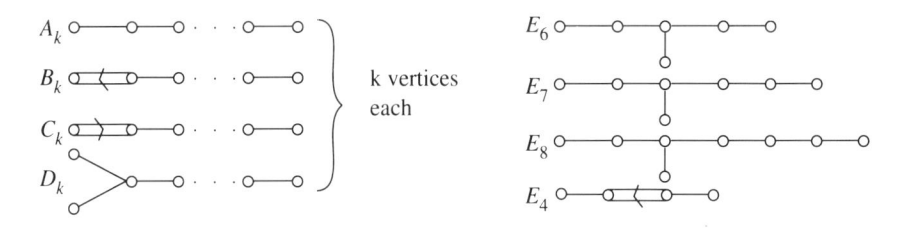

FIGURE 4

The diagrams B_k, C_k, and F_4 with double edges can be obtained from the action of nontrivial involutions on the canonical diagrams of the singularities A_{2k-1}, D_{k+1}, and E_6. The singularities in the ordinary nonsymmetric classification that correspond to the latter diagrams have the same normal forms as the symmetric singularities B_k, C_k, and F_4.

§2. Symmetric singularities

Let us imbed $\mathbb{C}^n$ into the space $\mathbb{C}^{n+1}$ with coordinates $z_0, \ldots, z_n$ as a plane $\mathbb{C}^n = \{z : z_0 + \cdots + z_n = 0\}$. The permutation group $S(n+1)$ of coordinates in $\mathbb{C}^{n+1}$ maps this plane into itself. Let $\lambda_1, \ldots, \lambda_n$ denote the coefficients of a polynomial $z^{n+1} + \lambda_1 z^{n-1} + \cdots + \lambda_n$ with zeroes $z_0, \ldots, z_n$. These functions form a basis in the space of symmetric (i.e., $S(n+1)$-invariant) functions on $\mathbb{C}^n$. The beginning of the hierarchy of symmetric functions on $\mathbb{C}^n$ near the critical point 0 looks as follows [5]:

$n+1$	Type	Normal form	m_f	m_h	codim
2	I_k	$\pm\lambda_1^k$	0	0	$k-1$
3	I	$\pm\lambda_1$	0	0	0
3	II	$\pm\lambda_2$	0	0	1
3	III_k	$\pm\lambda_1^2 + a\lambda_2^k$, $a \neq 0$	1	0	k
3	IV	$\pm\lambda_1\lambda_2 + a\lambda_1^3$	1	0	3
3	V		≥ 2	≥ 2	4
4	I	$\pm\lambda_1$	0	0	0
4	II	$\pm\lambda_2 + a\lambda_1^k$, $a \neq 0$	1	0	$k-1$
4	III	$\pm\lambda_2 + a\lambda_1^k$, $a \neq 0$	2	1	2
4	IV		≥ 2	≥ 2	3
5	I	$\pm\lambda_1$	0	0	0
5	II	$\pm\lambda_2 + a\lambda_1^2$	1	0	1
5	III		≥ 4	≥ 2	2

Here m_f and m_h are the numbers of moduli (i.e., of continuous invariants) of each function and its zero level hyperspace. The codimension in the space of function germs with critical value 0 at the critical point 0 refers to all singularities with the normal form indicated.

The adjacency diagrams are:

$$n = 2: \quad I_1 \leftarrow I_2 \leftarrow \cdots$$

$$n = 3: \quad I \leftarrow II \leftarrow III_2 \leftarrow III_3 \leftarrow \cdots$$
$$\uparrow$$
$$IV \leftarrow V$$

$$n = 4: \quad I \leftarrow II_2 \leftarrow II_3 \leftarrow \cdots$$
$$\uparrow$$
$$III \leftarrow IV$$

$$n = 5: \quad I \leftarrow II \leftarrow III$$

§3. General theorems

We have presented the first part of the hierarchies of invariant germs for an involution and for a symmetric group. The classification must be done separately for each symmetry group. The only existing general classification result concerns the very beginning of the hierarchy of critical points.

EQUIVARIANT MORSE LEMMA [5]. *Let G be a compact Lie group acting linearly on $\mathbb{C}^n$. Then the germ of any G-invariant holomorphic function that has a nondegenerate critical point at 0 with critical value 0 can be reduced to its quadratic part by a change of the independent variables which is G-equivariant (i.e., commutes with the action of G) and biholomorphic at zero.*

Similar theorems are valid in real-analytic and in C^∞ cases.

In classifying equivariant germs and analyzing their properties, it is useful to keep in mind that general finite determinacy and versality theorems can often be proved. We shall formulate these assertions in the holomorphic situation, although the C^∞ and the real-analytic analogues are also true.

Consider a finite graph that is the disjoint union of a finite number of trees. Assume that all its edges are oriented toward the roots (Figure 5). To each vertex α we associate a germ of linear space $(V_\alpha, 0)$ with a representation ρ_α of a compact or reductive Lie group G in that space. To each edge $\alpha \to \beta$ we assign a germ of a holomorphic map $f_{\alpha\beta} : (V_\alpha, 0) \to (V_\beta, 0)$ that commutes with the action of G: $\rho_\beta(g) f_{\alpha\beta} \rho_\alpha(g^{-1}) = f_{\alpha\beta}$ for all g in G. Let $\mathscr{F}$ be the space of all such collections $f = \{f_{\alpha\beta}\}$. The direct product $\mathscr{G}$ of the groups of germs of G-equivariant biholomorphisms h_α of spaces V_α acts on $\mathscr{F} : \{f_{\alpha\beta}\} \to \{h_\beta \circ f_{\alpha\beta} \circ h_\alpha^{-1}\}$.

DEFINITION. A collection of maps $f = \{f_{\alpha\beta}\}$ is called *finitely determined* if there is a finite number k, such that any other collection in $\mathscr{F}$ whose elements have the same k-jets at zero as the elements of f, is $\mathscr{G}$-equivalent to f.

Let T_f be the tangent space to the orbit $\mathscr{G}_f$ at f. The tangent space to $\mathscr{F}$ at any of its points is identified with $\mathscr{F}$.

THEOREM. *If $\dim_\mathbb{C} \mathscr{F}/T_f < \infty$, then the collection of maps f is finitely $\mathscr{G}$-determined.*

Consider a deformation $F : (\Lambda, 0) \to (\mathscr{F}, f)$ of a collection f. Let $\operatorname{Im} d_0 F$ be the image of its differential at zero.

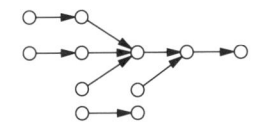

FIGURE 5

THEOREM. *F is $\mathscr{G}$-versal if and only if it is transversal to the G-orbit of f*, i.e., $\mathscr{F} = T_f + \operatorname{Im} d_0 F$.

REMARK. The above theorems cannot be extended to the case of C^∞ maps that are equivariant under the action of a reductive group.

For more general versions of the versality and finite determinacy theorems, see [7]. The validity of the above assertions follows from these more general statements and from the following description of invariant functions.

Let G be a compact or reductive Lie group, acting linearly on a finite-dimensional vector space V. The ring of polynomial invariants of this action is finitely generated. Choosing generators of this ring, we get a map $\varphi : V \to L$ into another finite-dimensional vector space.

THEOREM [15]. *The preimage $\mathscr{G}$ of the ring of germs of C^∞, real-analytic, or holomorphic functions on L coincides with the ring of G-invariant function germs on V of the appropriate class (if G is a reductive group, the C^∞ case is excluded).*

This theorem identifies G-invariant functions on V with functions on the image V/G of the factorization map φ. When can this identification be extended to equivalence classes of invariants and of functions on the quotient space up to diffeomorphisms of V/G?

We answer this question on the infinitesimal level, i.e., for Lie algebras of the corresponding vector fields. Incidentally, from a practical point of view it is more convenient to work with V/G than with invariants on V, since the calculations turn out to be less cumbersome.

If G is compact, the preimage of each point of V/G is exactly one orbit. If G is reductive, the preimage may contain several orbits, but only one of them is closed. The space V/G is *stratified by types of orbits*, that is, by the conjugacy classes in G of the stationary subgroups of points in V. It is easy to see that any G-symmetric vector field on V is mapped by φ onto a field on V/G, which preserves the above stratification (see, for example, the discussion of the correspondence between $\mathbb{Z}_2$-invariants and functions on a manifold with boundary in §1).

DEFINITION. A pair (G, V) has the *lifting property* if any field on V/G tangent to every stratum of the orbit decomposition of V/G has a φ-preimage that is a G-symmetric field on V.

It is shown in [16] that if the lifting property holds for algebraic fields, it holds for both C^∞ and analytic fields, and vice versa.

THEOREM [16]. *A pair (G, V) possesses the lifting property in the following cases:*

(1) *for every representation of a compact Lie group in a real space;*
(2) *for the orthogonal action of a reductive Lie group on a complex space;*
(3) *for the action of a finite group;*
(4) *if $\dim(V/G) = 1$.*

EXAMPLES. 1) Let G be a finite irreducible group generated by reflections acting on the complexification $\mathbb{C}^n$ of Euclidean space. The basis of G-symmetric fields on $\mathbb{C}^n$ consists of the gradients of basic invariants $\nabla \varphi_i$, $i = 1, \ldots, n$. Upon factorization they become the vector fields

$$\sum_{j=1}^{n} (\nabla \varphi_i, \nabla \varphi_j) \partial \varphi_j$$

on $\mathbb{C}^n/G \simeq \mathbb{C}^n$ (the parentheses denote the Euclidean scalar product).

2) In the following example the lifting property does not hold [18]. Consider the following action of the group $\mathbb{C}^*$ of nonzero complex numbers on $\mathbb{C}^3$:

$$t(x, y, z) = (t^a x, t^b y, t^{-c} z) \quad \text{where } a, b \geq 1 \text{ and } c \geq 2 \text{ are integers.}$$

The orbit space of this action can be identified with that of the action of the cyclic group of order c on $\mathbb{C}^2$:

$$(X, Y) \mapsto (\varepsilon^a X, \varepsilon^b Y), \qquad \varepsilon = \exp(2\pi i/c).$$

Invariant fields of the latter action are linear combinations of the fields

$$X^r Y^s \partial x, \qquad a(r-1) + bs = ck,$$
$$X^r Y^s \partial y, \qquad ar + b(s-1) = ck,$$

where in both cases $r, s \geq 0$ and k is an integer. One immediately sees that these fields lift to $\mathbb{C}^3$ only if $k \geq 0$. In that case, the field $X^r Y^s \partial x$, for example, has the corresponding $\mathbb{C}^*$-invariant field $x^r y^s z^k \partial x$ on $\mathbb{C}^3$. At the same time, it is easy to choose a, b, c such that the above relations also hold for a certain number $k < 0$ (the set of such numbers is finite).

§4. Equivariant Milnor numbers

Recall that a critical point 0 of a holomorphic germ $f : (\mathbb{C}^n, 0) \to (\mathbb{C}, 0)$ is isolated if and only if the local ring $\mathscr{O}_n/J(f)$ is finite-dimensional, where $J(f)$ is the ideal generated by the partial derivatives $\partial f/\partial x_i$. The dimension μ of $\mathscr{O}_n/J(f)$ is called the Milnor number (multiplicity) of the singularity. This number coincides with the rank of the middle dimension homology group of nonsingular level sets of f (the other homology groups are trivial). The number of Morse (nondegenerate) critical points into which a critical point 0 of f splits upon morsification (i.e., under a small generic perturbation) is also μ. In this section we consider equivariant analogues of all these numbers.

Let G be a finite group acting linearly on $\mathbb{C}^n$, $\mathscr{O}_n^G$ the ring of germs at 0 of its holomorphic invariants, f one such germ, and $J^G(f) = J(f) \cap \mathscr{O}_n^G$ the invariant part of the Jacobian ideal $J(f)$.

One immediately shows that a critical point 0 of a symmetric germ f is isolated if and only if the dimension $\nu = \dim_{\mathbb{C}} \mathscr{O}_n^G/J^G(f)$ of the base of a G-invariant miniversal deformation is finite, or if and only if the restriction f_g

of f to the stationary subspace V_g of any element g of G has an isolated singularity at zero.

The action of G on $\mathbb{C}^n$ induces a linear action on $\mathscr{O}_n : (g \cdot f)(x) = f(g \cdot x)$. This action maps the Jacobian ideal of an invariant function into itself. We may therefore introduce the following

DEFINITION. The *equivariant Milnor number* μ_G of a G-invariant f is the character of a representation of G in $\mathscr{O}_n/J(f)$.

Therefore, the ordinary Milnor number μ is $\mu_G(1)$. The number ν is given by $\nu = |G|^{-1} \sum_{g \in G} \mu_G(g)$.

Let X be a nonsingular fiber of f (local level $f = \varepsilon$, see subsection 4 in §1). The action of G on $\mathbb{C}^n$ also induces a linear action on the homology $H_{n-1}(X; \mathbb{C})$. Denote the character of this representation by ψ.

Let $\det(g)$ be the determinant of the operator $g \in G$ of the original representation of G on $\mathbb{C}^n$.

THEOREM [17]. *We have* $\psi(g) = \det(g) \cdot \mu_G(g)$.

DEFINITION. The character $\chi_G = 1 - (-1)^n \psi$ of a nonsingular fiber X is called the *equivariant Euler characteristic*.

THEOREM. *We have* $\chi_G(g) = 1 - (-1)^{n_g} \mu_g$, *where* $n_g = \dim V_g$, *and* μ_g *is the multiplicity of* f_g.

COROLLARY. *We have* $\psi(g) = (-1)^{n-n_g} \mu_g \cdot \mu_G(g) = (-1)^{n-n_g} \det^{-1}(g) \mu_g$.

If the original representation of G in $\mathbb{C}^n$ is the complexification of a real representation, then $\det(g) = (-1)^{n-n_g}$ and $\mu_G(g) = \mu_g$. In that case f also has a G-invariant morsification [14]. The action of G is to permute the set Ω of critical points of such a morsification. Let θ be the character of the induced linear representation of G on the set of mappings $\Omega \to \mathbb{C}$.

THEOREM [14]. *We have* $\Theta = \mu_G$.

It is easy to see that the number of G-orbits in Ω is ν. Let us see how these orbits break up into types, that is, into conjugacy classes of the isotropy subgroup of their points.

Let V_H be the stationary subspace of an isotropic subgroup H. Let K denote the set of all subspaces V_H in $\mathbb{C}$, where $H \subset G$ runs over all possible isotropy groups. We also denote:

$\eta : K \to \mathbb{Z}$ — the function $H \mapsto \mu(f_H)$ (multiplicity of the restriction of f to V_H);

$m : K \times K \to \mathbb{Z}$ — the Möbius function on the set K, ordered by inverse inclusion;

$N(H)$ — the normalizer of H in G;

$c(H)$ — the number of G-orbits in G whose isotropy subgroups are conjugate to H.

THEOREM [14]. *We have*

$$c(H) = \frac{|H|}{|N(H)|} \sum_{W \geq V_H} m(V_H, W) \cdot \eta(W).$$

EXAMPLES. 1) Suppose G acts on $\mathbb{C}^n$ as a group generated by real reflections. Let f_* be the function determined on $\mathbb{C}^n/G \simeq \mathbb{C}^n$ by a G-invariant f on $\mathbb{C}^n$. Obviously, $\mu(f_*) = c(1)$. According to [17], $\mu(f_*) = |G|^{-1} \sum_{g \in G} (-1)^{n-n_g} \mu_g$. The same expression can be obtained for $c(1)$ by using the formula of the last theorem together with certain properties of the Möbius function [13].

2) Let f be the function defined in §1, which is invariant under reflection σ in the hyperplane $x_1 = 0$; let f' be the restriction of f to this hyperplane. Then $\psi(\sigma) = \dim H^+ - \dim H^- = \mu(f')$. From the previous example we have $\mu(f_*) = (\mu(f) - \mu(f'))/2$. Since $\dim H^+ + \dim H^- = \mu(f)$, this implies $\dim H^+ = \mu(f_*)$, $\dim H^- = \mu(f_*) + \mu(f')$. This agrees with subsections 4 and 5 of §1: in the notation of §1 we have $\mu(f_+) = \nu_2$, $\mu(f') = \nu_1$.

§5. Singularity splitting

We will study the following questions. In which case will the deformation of an invariant function with a singular point at $0 \in \mathbb{C}^n$ produce a critical point of a given orbital type?

Let G be a compact Lie group acting linearly on $\mathbb{C}^n$, and $\mathbb{C}^n/G$ the quotient space, imbedded in a certain linear space $\mathbb{C}^m$. Let f be the germ at 0 of a G-invariant function on $\mathbb{C}^n$, and let f_* be a function on $\mathbb{C}^m$ whose preimage under factorization is f. For a subgroup H of G let $S_H \subset \mathbb{C}^m$ denote the stratum in $\mathbb{C}^n/G$ corresponding to orbits whose isotropy subgroups are conjugate to H.

THEOREM. *A critical point of type H can be obtained by deformation of a critical point 0 of a germ f if and only if the gradient of f_* in $T^*\mathbb{C}^m$ at 0 lies in the closure of the normal bundle of S_H.*

We now turn to the maximal isotropy conjecture.

Let G be a compact Lie group that acts absolutely irreducibly on $\mathbb{R}^n$. We consider G-equivariant maps $\mathbb{R}^n \to \mathbb{R}^n$ (with the same representation in the image and in the preimage).

CONJECTURE [9]. *Let $F : \mathbb{R}^n \times \mathbb{R} \to \mathbb{R}^n$, $F(0, 0) = 0$, be a generic one-parameter family of G-equivariant maps. Then every nontrivial branch of a solution of the equation $F = 0$ has a real maximal isotropy subgroup. Moreover, every such subgroup is realized in some one-parameter family in general position.*

Golubitsky also conjectured that this result remains true if all real maximal isotropy subgroups have one-dimensional stationary subspaces.

The conjecture was verified in [8] for the Weyl groups A_k, B_k, F_4, $I_2(p)$, and H_3. However, it fails for D_k, $k \geq 4$. The point is that the stratification of the orbit space of D_k includes a two-dimensional stratum whose fiber over 0 in the closure of the normal bundle is a hyperplane (not a plane of codimension 2). The conjecture also fails for representations of $SO(3)$ and $O(3)$ in high dimensions [6, 12]. For some other questions connecting the symmetry modification, see [10, 11].

§6. Complete models in physics

In this section we present several applications to the theory of phase transitions in physical systems with symmetries. We recall the definition of a phase transition of kth order.

Let us assume that some physical system is described in terms of a generic family of potential functions $F(x, \lambda)$, depending on n state variables $x \in \mathbb{R}^n$ and ℓ control parameters $\lambda \in \mathbb{R}^\ell$. Suppose further that the state of the system is described by a value of x that minimizes the potential function, at least locally. Assigning to each value of the control parameter λ the local minima $x^{(p)}(\lambda)$, $p = 1, \ldots$, we obtain a many-valued map $\lambda \to x$. A phase transition occurs at the moment when the point $x \in \mathbb{R}^n$ describing the state of the system jumps from one branch of this map to another.

Imbedded in the λ-space we have the bifurcation diagram Ξ of the functions, i.e., the set of λ's at which $F(\cdot, \lambda)$ has either degenerate critical points or fewer different critical values than a function of the same family with generic values of control parameters close to λ. In a real process one usually assumes that control parameters themselves depend on just one parameter, say, the time t. This yields a curve $\lambda(t)$ in $\mathbb{R}^\ell$. Depending on which physical principle we adopt, phase transitions occur when the curve $\lambda(t)$ intersects some component of the diagram Ξ. If we adopt the so-called *maximal delay principle*, the phase transitions will take place on components corresponding to the appearance and disappearance of local minima (Figure 6a). On the other hand, *Maxwell's principle* implies their occurrence on the components corresponding to two global minima of the potential (Figure 6b).

It is convenient to use Ehrenfest's classification of phase transitions. Let $\varphi : t \mapsto F(x(\lambda(t)), \lambda(t))$ be the function that associates to each point of the curve $\lambda(t)$ the corresponding minimal value of the potential. We say that a

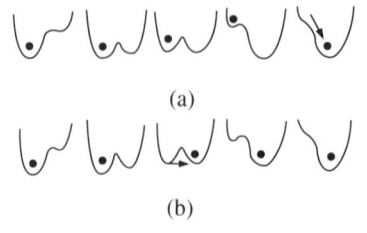

(a)

(b)

FIGURE 6

FIGURE 7. $F(x, t) = x^4 + tx^2$

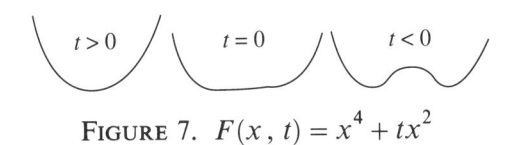

phase transition of order k occurs at time t_0 if φ is exactly $k-1$ times continuously differentiable at t_0.

Phase transitions in actual physical systems are usually of order zero (Figure 6a), one (Figure 6b), or two. Transitions of arbitrary order can occur only in systems with symmetry, and, contrary to transitions of zero or first order, they are local rather than profound in nature (Figure 7).

In physics we often encounter the following problem. Suppose it is known from experiment that a certain set of phenomena occurs in a given physical system when the control parameters are varied (states of different symmetries and phase transitions between them are observed, and so on). It is required to choose a potential function of the simplest possible form to describe the system.

Let us call a potential function a *complete model* for a fixed set of phenomena if it has a versal deformation with these phenomena taking place in the base. We say that a given complete model is *minimal* if it does not adjoin any other complete model (for the same set of phenomena).

Let G be a finite irreducible group generated by reflections in $\mathbb{R}^n$, f an invariant of G, f_* the corresponding germ in the quotient space, and i_2 a basic second-order G-invariant.

THEOREM. (a) *A function f, $f(0) = 0$, is a minimal complete model for the set of states of all possible symmetry types, and also for all possible second-order phase transitions between states of different symmetry types, if and only if the 2-jet of f_* at zero is a nondegenerate quadratic form in the basis invariants.*

(b) *A function f is a minimal complete model for all possible first- or zero-order phase transitions between states with different symmetry types if and only if f_* has a critical point at zero of the class A_3^+ in the direction of i_2.*

The last condition has the following meaning. f_* is a function on the space $\mathbb{R}^n$ in which the quotient space $\mathbb{R}^n/G$ is imbedded. Its linear part at zero should be trivial, the kernel of its quadratic part must be generated by the vector ∂i_2, and the condition $\partial^3 f_*/\partial i_2^3(0) = 0$ should be satisfied. Moreover, there is a diffeomorphism of $\mathbb{R}^n$, not preserving the imbedded quotient $\mathbb{R}^n/G$, that transforms the germ f_* at zero to the form $y_1^4 + y_2^2 + \cdots + y_n^2$.

References

1. V. I. Arnol'd, *Critical points of function on a manifold with boundary, the simple Lie groups B_k, C_k, F_4 and singularities of evolutes*, Uspekhi Mat. Nauk **33** (1978), no. 5, 91–105; English transl. Russian Math. Surveys **33** (1978), no. 5, 99–116.
2. E. Brieskorn, *Sur les groupes de tresses (d'après V. I. Arnol'd)*, Sém. Bourbaki (1971/72), Exp. 401, Lecture Notes in Math., vol. 317, Springer-Verlag, Berlin, 1973.
3. N. Bourbaki, *Groupes et algèbres de Lie*. VII, Hermann, Paris, 1968; Masson, Paris, 1981.
4. O. V. Lyashko, *The geometry of bifurcation diagrams*, Uspekhi Mat. Nauk **34** (1979), no. 3, 205–206; English transl. in Russian Math. Surveys **34** (1979), no. 3.
5. V. I. Arnol'd, *Wave front evolution and equivariant Morse lemma*, Comm. Pure Appl. Math. **29** (1976), no. 6, 557–582.
6. P. Chossat, *Solutions avec symetrie diedrale dans le problème de bifurcation invariants par symetrie spherique*, C. R. Acad. Sci. Paris Sér. I Math. **27** (1983), 639–642.
7. J. Damon, *The unfolding and determinacy theorems for subgroups of A and K*, Mem. Amer. Math. Soc. **50** (1984), no. 306.
8. M. J. Field and R. W. Richardson, *Symmetry breaking and the maximal isotropy subgroup conjecture for reflection groups*, Arch. Rational Math. Anal. **105** (1989), 61–94.
9. M. Golubitsky, *The Benard problem, symmetry and the lattice of isotropy subgroups*, Bifurcation Theory, Mechanics and Physics (C. P. Bruter et al., eds.), Reidel, Dordrecht, Boston, Lancaster, 1987, pp. 225–257.
10. M. Golubitsky and D. G. Schaeffer, *Singularities and groups in bifurcation theory*. I, Applied Math. Sciences, vol. 51, Springer-Verlag, Berlin, New York, 1985.
11. M. Golubitsky, I. Stewart, and D. G. Schaeffer, *Singularities and groups in bifurcation theory*. II, Applied Math. Sciences, vol. 69, Springer-Verlag, New York, 1988.
12. R. Lauterbach, *An example of symmetry breaking with submaximal isotropy subgroup*, Multiparameter Bifurcation Theory (Arcata, Calif., 1985), Contemp. Math., vol. 56, Amer. Math. Soc., Providence, R. I., 1986, pp. 217–222.
13. P. Orlik and L. Solomon, *Unitary reflection groups and cohomology*, Invent. Math. **59** (1980), 77–94.
14. M. Roberts, *Equivariant Milnor numbers and invariant Morse approximations*, J. London Math. Soc. **31** (1985), 487–500.
15. G. W. Schwarz, *Smooth functions invariant under the action of a compact Lie group*, Topology **14** (1975), 63–68.
16. _____, *Lifting smooth homotopies of orbit spaces*, Inst. Hautes Études Sci. Publ. Math. **51** (1980), 37–135.
17. C. T. C. Wall, *A note on symmetry of singularities*, Bull. London Math. Soc. **12** (1980), 169–175.
18. _____, *Functions on quotient singularities*, Philos. Trans. Roy. Soc. London Ser. A **324** (1982), no. 1576, 1–45.

Translated by L. G. HANIN

Stratifications of Function Space and Algebraic K-theory

S. M. GUSEĬN-ZADE

The purpose of this paper is to examine the relationship between the methods of singularity theory and algebraic K-theory in terms of the problem of computing homotopy groups of spaces of pseudo-isotopies. This problem is closely related to the analogous problem for spaces of diffeomorphisms. Our discussion will be based on work done by Volodin, Wagoner, and Hatcher in the early 1970s; more recent progress in the theory will not be touched upon. Arnol'd has suggested an interesting problem of constructing a complex analogue of that theory, but so far it is not clear how to approach it.

Since the detailed exposition would involve several complicated technical tools, we shall sometimes intentionally sacrifice not only consistency but also rigor concentrating primarily on those constructions whose description fits our purpose. We will basically try to follow the line of Volodin's arguments (not yet fully published).

Let us first define (or recall) some elementary notions of algebraic K-theory. Given an algebra $\mathscr{A}$ (associative with unit), one defines certain groups $K_i(\mathscr{A})$, $i = 0, 1, \ldots$. When these concepts were first defined, the main requirement was that, if $\mathscr{A}$ is the algebra of (say, complex-valued) functions on a topological space X (which is of course assumed to be sufficiently "well behaved," e.g., CW-complex), one should obtain the usual topological K-functor of X.

In topology, $K^0(X) = K(X)$ is the group of (complex) stable vector bundles over X, i.e., equivalence classes of vector bundles with respect to stable equivalence, with addition induced by the direct sum operation for bundles. Two vector bundles are *stably equivalent* if they become isomorphic after adding trivial bundles of certain dimensions. The groups $K^{-i}(X)$ are defined as $K(\Sigma^i X)$, where $\Sigma^i X$ is the ith suspension over X. A *suspension* over X is, in a sense, a one-parameter family of X's beginning and ending at a point. An element of $K^{-1}(X)$ is determined by a vector bundle over $\Sigma^i X$. It can be thought of as a one-parameter family of bundles over X (i.e., representatives of elements of $K^0(X)$) which both begins and ends at a trivial object (with fixed trivialization!). Thus, an element of $K^{-1}(X)$ can be viewed as a process of matching two trivial objects of $K^0(X)$. An element of $K^{-2}(X)$ (i.e., of $K(\Sigma^2 X)$) can be viewed as a one-parameter family of elements of $K^{-1}(X)$, that is, a process of matching two trivial objects of $K^{-1}(X)$; and so on.

Let us see how to imitate this description in the algebraic situation. In algebra, one usually writes K_i for K^{-i}.

Let $\mathscr{A}$ be an algebra (associative, with unit). In the sequel, $\mathscr{A}$ will be a group algebra $Z[\pi]$, where π is a group, in general noncommutative (the fundamental group of a certain manifold). Elements of $Z[\pi]$ are finite formal sums $\sum n_i g_i$ ($g_i \in \pi$, n_i integers) with the natural operations of addition and multiplication. Guided by the above requirements and similarities, one can suggest definitions of the groups $K_i(\mathscr{A})$ (see [3]). We present these definitions for $i = 0, 1, 2$ only (the others will not be needed).

1) $K^0(\mathscr{A})$. A vector bundle over a space X determines a projective module over the ring of functions on X, the module of its sections. Conversely, the module of sections determines the bundle. Addition of a trivial bundle to a given bundle enlarges its module of section by a free module over the ring of functions. Therefore, $K_0(\mathscr{A})$ is defined as the group of stable-equivalence classes of projective $\mathscr{A}$-modules, where two modules are said to be *stably equivalent* if they become isomorphic after the addition of trivial, i.e., free $\mathscr{A}$-modules. The trivial object (zero) is that by the free $\mathscr{A}$-modules.

2) $K_1(\mathscr{A})$ and $\text{Wh}(\pi)$. A process of matching two trivial objects in $K_0(A)$ with fixed trivializations, i.e., two free $\mathscr{A}$-modules with fixed $\mathscr{A}$-bases, is described by an invertible matrix with elements in $\mathscr{A}$, that is, an element of $\text{GL}(\mathscr{A}) = \text{GL}(\mathscr{A}, \infty) = \lim \text{GL}(\mathscr{A}, n)$. Therefore, $K_1(\mathscr{A})$ is the maximal commutative group obtained by passing from $\text{GL}(\mathscr{A})$ to its quotient groups, i.e., $K_1(\mathscr{A}) = \text{GL}(\mathscr{A})/E(\mathscr{A})$, where $E(\mathscr{A}) = [\text{GL}(\mathscr{A}), \text{GL}(\mathscr{A})]$ is the commutator subgroup of $\text{GL}(\mathscr{A})$.

It can be shown that $E(\mathscr{A})$ is generated by the matrices $\alpha_{ij}(a) = E + \Delta_{ij}(a)$, where E is the identity matrix and $\Delta_{ij}(a)$ the matrix with a single nonzero element a in the (ij) position. Thus $K_1(\mathscr{A})$ can be defined as the group of equivalence classes of matrices of arbitrary sizes with elements in $\mathscr{A}$, where the following pairs of matrices are defined to be equivalent:

(1) X and $\begin{pmatrix} X & 0 \\ 0 & E \end{pmatrix}$, where E is any identity matrix;
(2) any two matrices obtained from one another by permutations of rows or columns;
(3) any two matrices obtained from one another by adding a multiple of a row (or column) to another row (or column).

Starting from this definition, one can show that a superposition XY (product of matrices) coincides in K_1 with the Whitney sum $\begin{pmatrix} X & 0 \\ 0 & Y \end{pmatrix}$ (see [2]).

If $\mathscr{A} = Z[\pi]$, $K_1(\mathscr{A})$ contains the subgroup generated by the elements $\pm \pi$ (i.e., by 1×1 matrices). The quotient group $K_1 / \pm \pi$ is called the *Whitehead group* of π, and denoted by $\mathrm{Wh}(\pi)$.

An element of $\mathrm{Wh}(\pi)$ that corresponds to a matrix H is a natural generalization of the determinant concept to matrices with elements in $Z[\pi]$. It is often convenient to define an element of $\mathrm{Wh}(\pi)$ not only for a single isomorphism H of free $Z[\pi]$-modules (with fixed bases), i.e., for a short exact sequence $0 \leftarrow (Z[\pi])^K \overset{H}{\leftarrow} (Z[\pi])^K \leftarrow 0$, but for a long exact sequence of free $Z[\pi]$-modules:

$$0 \leftarrow \mathscr{M}_0 \leftarrow \mathscr{M}_1 \leftarrow \mathscr{M}_2 \leftarrow \cdots \leftarrow \mathscr{M}_n \leftarrow 0.$$

One can consider the set of equivalence classes of long exact sequences with respect to the equivalence relation defined by

(1) permuting basis elements in any module $\mathscr{M}_i$;
(2) adding a multiple of a basis element in a module $\mathscr{M}_i$ to another basis element;
(3) adding or subtracting direct free summands, in the form of short exact sequences

$$0 \leftarrow (Z[\pi])^K \overset{\mathrm{id}}{\leftarrow} (Z[\pi])^K \leftarrow 0,$$

at any position in the sequence;
(4) multiplying elements in the basis of any $\mathscr{M}_i$ by elements from $\pm \pi$.

Addition of exact sequences is defined in the natural way. It can be shown that the group thus obtained is just the Whitehead group $\mathrm{Wh}(\pi)$ [2]. In what follows we shall not indicate precisely whether we are dealing with arbitrary or only short exact sequences. In differential topology the operation of "telescoping" arbitrary exact sequences into short ones corresponds to eliminating all the critical points of a function with nondegenerate critical points on a manifold except for a pair of adjacent indices i and $i+1$.

3. $K_2(\mathscr{A})$ and $\mathrm{Wh}_2(\pi)$. A matrix in $\mathrm{GL}(\mathscr{A})$ determines an element of $K_1(\mathscr{A})$, which is zero if it lies in the subgroup $E(\mathscr{A})$, i.e., if it can be expressed as a product of matrices $\alpha_{ij}(a)$. Hence any matching of two trivial objects means representing an identity matrix as a product $\prod_S \alpha_{i_s j_s}(a_s)$. We define the Steriberg group $\mathrm{St}(\mathscr{A})$ as the quotient group of the free group with formal generators $\alpha_{ij}(a)$ $(a \in \mathscr{A})$ modulo the universal relations

$$\alpha_{ij}(a+b) = \alpha_{ij}(a) \cdot \alpha_{ij}(b),$$
$$[\alpha_{ij}(a)\alpha_{k\ell}(b)] = 1, \quad i \neq \ell, \ j \neq k,$$
$$[\alpha_{ij}(a)\alpha_{jk}(b)] = \alpha_{ik}(ab)$$

(the above matrices $\alpha_{ij}(a)$ satisfy these relations). The group $K_2(\mathscr{A})$ is

defined as the kernel of the natural epimorphism $\mathrm{St}(\mathscr{A}) \to E(\mathscr{A})$:
$$0 \to K_2(\mathscr{A}) \to \mathrm{St}(\mathscr{A}) \to E(\mathscr{A}) \to 0.$$
Let $\mathscr{A} = Z[\pi]$, and let $W \subset \mathrm{St}(Z[\pi])$ be the subgroup generated by the elements $w_{ij}(\pm g) = \alpha_{ij}(\pm g)\alpha_{ji}(\mp g^{-1})\alpha_{ij}(\pm g)$, $g \in \pi$. The epimorphism $\mathrm{St}(Z[\pi]) \to E(Z[\pi])$ maps an element $w_{ij}(\pm g)$ to a matrix that differs from the identity matrix only in the block corresponding to the ith and jth basis elements; this block is $\begin{pmatrix} 0 & \pm g \\ \mp g^{-1} & 0 \end{pmatrix}$. Thus the elements $w_{ij}(\pm g)$ correspond to permutations of the basis elements, appropriately multiplied by elements $\pm \pi$. Let $W_0 = K_2(Z[\pi]) \cap W)$. The quotient group $K_2(Z[\pi])/W_0$ is called the *second Whitehead group* $\mathrm{Wh}_2(\pi)$ of π.

Going on now to the realm of differential topology, let $\mathscr{M} = \mathscr{M}^n$ be a smooth compact manifold. To fix ideas, we assume that $\mathscr{M}$ has no boundary (for manifolds with boundary, some of the subsequent definitions must be modified).

An isotopy of $\mathscr{M}$ is a one-parameter family of diffeomorphisms $h_t: \mathscr{M} \to \mathscr{M}$ with $h_0 = \mathrm{id}$ or, in other words, a diffeomorphism $h: \mathscr{M} \times I \to \mathscr{M} \times I$ ($I = [0, 1]$) such that 1) $p_2 \cdot h = p_2$, 2) $h_{|\mathscr{M} \times 0} = \mathrm{id} \times 0$ ($p_2: \mathscr{M} \times I \to I$ is a projection onto the second factor). If we drop the first of these conditions, we get the notion of a pseudo-isotopy: a pseudo-isotopy of $\mathscr{M}$ is a diffeomorphism $h: \mathscr{M} \times I \to \mathscr{M} \times I$ of the cylinder $\mathscr{M} \times I$ that maps the bases $\mathscr{M} \times 0$ and $\mathscr{M} \times 1$ into themselves and is the identity on the lower base $\mathscr{M} \times 0$. Pseudo-isotopies of $\mathscr{M}$ form a space (in fact, a topological group) $P_i(\mathscr{M})$.

Description of the homotopy type of the space of pseudo-isotopies $P_i(\mathscr{M})$ can be reduced to the description of a certain subspace of functions. With any pseudo-isotopy $h: \mathscr{M} \times I \to \mathscr{M} \times I$ we associate the function $p_2 h$ on the cylinder $\mathscr{M} \times I$. This function has no critical points on $\mathscr{M} \times I$ and takes values 0 and 1 on the lower and the upper bases, respectively. It is easy to see that any function f satisfying these conditions can be obtained from some pseudo-isotopy (a mapping $\mathscr{M} \times I \to \mathscr{M} \times I$ corresponding to f can be defined, for example, by transforming the generators $m \times I$ of the cylinder $\mathscr{M} \times I$ ($m \in \mathscr{M}$) into the integral lines of the dynamical system $\mathrm{grad}\, f / \|\mathrm{grad}\, f\|^2$). We have thus defined a mapping of the space of pseudo-isotopies $p_i(\mathscr{M})$ onto the space $\mathscr{F}(0)$ of all functions $(\mathscr{M} \times I, \mathscr{M} \times 0, \mathscr{M} \times 1) \to (I, 0, 1)$ without critical points. It is easy to see that this mapping is a fibration (in the sense of Serre, for instance). Its kernel (fiber) is the set of all diffeomorphisms of the cylinder $\mathscr{M} \times I$ that preserve p_2, i.e., the semigroup of isotopies H. One immediately sees that H is contractible; hence the space of pseudo-isotopies $p_i(\mathscr{M})$ is homotopy equivalent to the space $\mathscr{F}(0)$ of functions on $\mathscr{M} \times I$ without critical points.

We may assume without loss of generality that we are considering only pseudo-isotopies that are constant near the lower and the upper bases or, accordingly, only functions on $\mathscr{M} \times I$ that coincide with p_2 in neighborhoods of the bases.

Let $\mathscr{F}$ be the space of all (C^∞) functions on $\mathscr{M} \times I$ that coincide with p_2 in neighborhoods of the bases. Of course, this space is contractible. Hence it can be derived from the exact sequence for the pair $(\mathscr{F}, \mathscr{F}(0))$ that $\pi_i(P_i(\mathscr{M})) \cong \pi_i(\mathscr{F}(0)) \cong \pi_{i+1}(\mathscr{F}, \mathscr{F}(0))$. Moreover, we may assume that $\mathscr{F}$ is the space of functions with only isolated critical points, or even only critical points with finite multiplicity. Functions with nonisolated critical points form a subset of infinite codimension in the space of all C^∞ functions, so its removal has no effect on the homotopy groups.

Our interest is precisely in the homotopy groups $\pi_{i+1}(\mathscr{F}, \mathscr{F}(0))$. However, we shall not go far in this direction, considering only *the groups* π_0 and π_1, which correspond, in a sense, to "π_{-1}" and π_0 for the space of pseudo-isotopies. Here $\pi_0(\mathscr{F}, \mathscr{F}(0))$ is, of course, trivial (since it is a set), and it is meaningless in this situation to speak of computing it. Nevertheless, in order to maintain our previous scheme, we will in fact deal with its computation, but in a somewhat different, perfectly meaningful situation: instead of functions on a cylinder $\mathscr{M} \times I$, we shall consider functions on an h-cobordism. An h-cobordism has approximately the same relation to a cylinder $\mathscr{M} \times I$ as a pseudo-isotopy has to isotopy.

An *h-cobordism* (with base $\mathscr{M}^n$) is a manifold W^{n+1} whose boundary is the union of two components $\mathscr{M}$ and $\widetilde{\mathscr{M}}$, each being a deformation retract of W. As $\mathscr{F}$ we must take the space of functions $(W, \mathscr{M}, \widetilde{\mathscr{M}}) \to (I, 0, 1)$ that are regular (i.e., have no critical points) in a neighborhood of $\partial W = \mathscr{M} \cup \widetilde{\mathscr{M}}$. Without loss of generality, we may confine attention to functions that coincide in a neighborhood of ∂W with some fixed function. To compute $\pi_0(\mathscr{F}, \mathscr{F}(0))$ means to determine whether the subspace $\mathscr{F}(0)$ of functions without critical points is empty.

The subspace $\mathscr{F}(0)$ is nonempty if and only if the h-cobordism $(W, \mathscr{M}, \widetilde{\mathscr{M}})$ is diffeomorphic to the cylinder $(\mathscr{M} \times I, \mathscr{M} \times 0, \mathscr{M} \times 1)$ (and then, in particular, $\mathscr{M}$ and $\widetilde{\mathscr{M}}$ are also diffeomorphic).

How can $\pi_i(\mathscr{F}, \mathscr{F}(0))$ be estimated? Assume that we have constructed a stratification of $\mathscr{F}$, i.e., a representation of $\mathscr{F}$ as a union of strata, each being a submanifold of finite codimension (with appropriate regularity conditions fulfilled for adjacent strata). Let $\mathscr{F}_i$ denote the skeleton of codimension i, i.e., the union of codimension $\geq i$ strata ($\mathscr{F}_0 = \mathscr{F} \supset \mathscr{F}_1 \supset \mathscr{F}_2 \supset \cdots$). Let $\mathscr{F}(0)$ be the union of the components of the complement of $\mathscr{F}_1$ in $\mathscr{F}$.

Suppose that some invariant, which is trivial for $\mathscr{F}(0)$, is associated with each component of $\mathscr{F} - \mathscr{F}_1$. We wish to calculate $\pi_0(\mathscr{F}, \mathscr{F}(0))$, that is, to find out whether every point of $\mathscr{F}$ (which we may assume without loss of generality to belong to $\mathscr{F} - \mathscr{F}_1$) can be connected by a path to a point in $\mathscr{F}(0)$. It will suffice to consider only paths not intersecting $\mathscr{F}_2$. Suppose we already know how to describe what happens to our invariant when the path intersects a codimension-one stratum (transversally). Suppose that intersection of the Kth stratum causes a transformation α_k of the invariant. Then a necessary condition for existence of a path connecting the point (function) in consideration in $\mathscr{F}$ to $\mathscr{F}(0)$ is that the invariant corresponding to the function be reducible to the trivial invariant by a finite sequence of transformations α_k.

Suppose we already know that $\mathscr{F}(0)$ is not empty (as is the case for functions on $\mathscr{M} \times I$), and let us calculate $\pi_1(\mathscr{F}, \mathscr{F}(0))$. Consider a path in $\mathscr{F}$ beginning and ending in $\mathscr{F}(0)$; we may assume that it does not intersect the skeleton $\mathscr{F}_2$. As the path intersects codimension-one strata in order of their appearance, we obtain a certain product of their formal elements α_k, which must result in the identical transformation of the invariant. A product of elements α_k is associated with every codimension-two stratum $S^{(2)}$ in the following way. Consider a two-dimensional area element transversal to $S^{(2)}$. The traces left by codimension-one strata on this element are curves beginning at the point corresponding to $S^{(2)}$ (see Figure 1). A small circle about this point is cut by these curves, i.e., by the codimension-one strata, in a certain order. It is this sequence of intersections that determines the product of α_k's. A necessary condition for a loop to determine the zero element of $\pi_1(\mathscr{F}, \mathscr{F}(0))$, i.e., to be homotopic to a point, is that the corresponding product of α_k's can be expressed as a product of relations determined by products associated with codimension-two strata. There are evidently certain similarities between this construction and the definition of K_2 or Wh_2; below they will be described in more detail.

In the general case the above construction yields an estimate of the homotopy groups which is neither an upper nor a lower bound. Some nontrivial elements of $\pi_1(\mathscr{F}, \mathscr{F}(0))$ may "escape" (for example, if the invariant is not chosen properly). Different strata may give algebraically equivalent transformations. In that case, naturally, reduction in the class of transformations cannot always be realized as a path homotopy. Finally, it may be impossible a priori to obtain certain elements (for example, products of generators α_k corresponding to identical transformations of the invariant). That is why effective realization of this construction requires proper choice of the stratification and of the initial invariant.

Stratifications of function spaces arise in a natural way in singularity theory. One stratification is that based on types of critical points. The functions with Morse (nondegenerate) critical points form an open dense subset of the space of smooth functions. Other functions are classified according to the codimensions of their critical points. Codimension-one functions occur in

FIGURE 1

an unremovable way (i.e., they cannot be eliminated by small displacements) in one-parameter families, those of codimension two, in two-parameter families, etc. Besides Morse critical points, functions with singularities of codimension one have exactly one critical point of type A_2 (i.e., equivalent to $x_0^3 + \sum_{i=1}^n \pm x_i^2$); codimension-two functions may have critical points of type A_3 (equivalent to $\pm x_0^4 + \sum_{i=1}^n \pm x_i^2$), and so on. However, this stratification falls short of satisfying our requirements and we must refine it (at the same time modifying the space $\mathscr{F}$ itself).

Let f be a function on an h-cobordism W (in particular, it may be a function on the cylinder $\mathscr{M} \times I$), which has only Morse critical points. As we know, for such functions we can construct a decomposition of the pair $(W, \mathscr{M})$ into handles, and the corresponding complex of (say, integral) chains:

$$0 \leftarrow C_0 \leftarrow C_1 \leftarrow C_2 \leftarrow \cdots \leftarrow C_n \leftarrow C_{n+1} \leftarrow 0.$$

The homology of this complex coincides with that of the pair $(W, \mathscr{M})$, and is therefore trivial. If $\mathscr{M}$ has a nontrivial fundamental group (this is the case of interest here), a similar chain complex can be constructed not only for the pair $(W, \mathscr{M})$ and the function f, but also for the universal covering $(\widehat{W}, \widehat{\mathscr{M}})$ and the function $\widehat{f} = f \circ p$, where $p : (\widehat{W}, \widehat{\mathscr{M}}) \to (W, \mathscr{M})$ is a projection. The pair $(\widehat{W}, \widehat{\mathscr{M}})$ also has trivial homology, and hence the corresponding sequence

$$0 \leftarrow C_0 \leftarrow C_1 \leftarrow C_2 \leftarrow \cdots \leftarrow C_{n+1} \leftarrow 0$$

is exact. In addition, the fundamental group $\pi = \pi_1(\mathscr{M}) = \pi_1(W)$ acts freely on the universal covering $(\widehat{W}, \widehat{\mathscr{M}})$. The elements of π permute the critical points of the function $\widehat{f}$, to which generators of the free abelian groups C_λ correspond. Hence the C_λ's are free $Z[\pi]$-modules, whose basis elements correspond to critical points of index λ of f. We wish to treat the exact sequence

$$\{C_\lambda\} : 0 \leftarrow C_0 \leftarrow C_1 \leftarrow \cdots \leftarrow C_{n+1} \leftarrow 0$$

of free $Z[\pi]$-modules with fixed bases, determined by f on $(W, \mathscr{M})$, as an invariant of the component of $\mathscr{F} - \mathscr{F}_1$. For functions in $\mathscr{F}(0)$ (i.e., without critical points) the above sequence clearly degenerates into the trivial sequence. However, when the space is stratified according to types of critical points, the exact sequence $\{C_\lambda\}$ may vary within the same component of $\mathscr{F} - \mathscr{F}_1$. In order to understand what modifications are required, we must recall the construction of relative chains corresponding to critical points of index λ (the generators of C_λ) for a function f on $(W, \mathscr{M})$ (or for a function $\widehat{f}$ on $(\widehat{W}, \widehat{\mathscr{M}})$) with Morse critical points. They can be represented geometrically as disks of dimension λ swept out by separatrices of the gradient vector field of f along which f decreases and which start at critical points. Corresponding to a critical point and a disk of the above type, on the universal covering $\widehat{W}$ we have a set of critical points of $\widehat{f}$ and a set of disks, which are freely permuted by the elements of π. The elements of $Z[\pi]$-matrices that correspond to the boundary homomorphisms $\partial_{\lambda+1}: C_{\lambda+1} \to C_\lambda$ can be described as follows.

Let $\Delta^{(\lambda+1)}$ and $\Delta^{(\lambda)}$ be basis cycles in $C_{\lambda+1}$ and C_λ, corresponding to (nondegenerate) critical points $a^{(\lambda+1)}$ and $a^{(\lambda)}$ of $\widehat{f}$ on $\widehat{W}$, of indices $\lambda + 1$ and λ, respectively. If $f(a^{(\lambda+1)}) \leq f(a^{(\lambda)})$, the coefficient of $\Delta^{(\lambda)}$ in $\partial_{\lambda+1}\Delta^{(\lambda+1)}$ (the incidence coefficient) is zero. If $f(a^{(\lambda+1)}) > f(a^{(\lambda)})$, then, in order to determine the coefficient, we have to consider the separatrices of the gradient dynamical system that go from $a^{(\lambda+1)}$ to all points of the orbit $\pi a^{(\lambda)}$ of $a^{(\lambda)}$. The necessary coefficient is $\sum n(g)g$, where $n(g)$ is the number of separatrices linking $a^{(\lambda+1)}$ to $ga^{(\lambda)}$, counting multiplicities.

With this definition, in order to construct disks corresponding to basis chains (i.e., elements of the modules C_λ), we need, in addition to the function f, a metric on W that determines the gradient vector field associated with f. Bifurcations of the family of integral curves of the gradient dynamical system (without any change in the number and types of the critical points of the function itself) may lead to a rearrangement of the system of basis elements. Hence, in order to describe the possible bifurcations, it becomes necessary to extend the fundamental space $\mathscr{F}$. A natural inclination is to replace the space of functions by its product with the space of Riemannian metrics on W. Since the set of Riemannian metrics is convex in the corresponding space and is therefore contractible, multiplication by this set does not change the homotopy type. However, it becomes substantially more complicated to describe the possible bifurcations if only *gradient* vector fields are used to define the generators of the modules C_λ corresponding to the critical points of f. The point is that, in a certain sense, there are not enough metrics—or, therefore, gradient vector fields—on a manifold. Deformations of gradient vector fields form a small part of all possible deformations, hence the list of families in general position for gradient vector fields may differ from that for arbitrary vector fields (the latter being much easier to obtain).

This is the reason why we consider what is called gradient-like vector fields instead of gradient vector fields. A C^∞ smooth vector field X on a manifold W is said to be *gradient-like* for a function f if it is a gradient field (with respect to some Riemannian metric) in small neighborhoods of the critical points of f, and everywhere outside these neighborhoods the derivative of f in the direction of X is positive. Any sufficiently small perturbation of a gradient-like vector field outside neighborhoods of critical points is a gradient-like vector field. For a function f with isolated critical points, the set of gradient-like vector fields is contractible; hence extension of a space of functions by the gradient-like vector fields associated with the functions does not change its homotopy type. After this extension, an element of $\mathscr{F}$ (a function f together with a gradient-like vector field X) must be regarded as degenerate (i.e., an element of the skeleton F_1) not only if f has non-Morse points, but also if the family of separatrices of X is not in general position (see below). The corresponding stratification of $\mathscr{F}$ should be based on the codimensions of the critical points of f and of the singularities of families of separatrices of a gradient-like vector field. In the sequel, it will often be convenient to speak of functions rather than pairs (function, gradient-like vector field).

The exact sequence $\{C_\lambda\}$ for a specific element of $\mathscr{F} - \mathscr{F}_1$, i.e., a function with Morse critical points and the corresponding gradient-like vector field whose families of separatrices are in general position, is determined almost uniquely. The ambiguity lies in the choice of 1) the critical point of $\widehat{f}$ among those projecting to the same critical point of f on W; and 2) the orientation of the disk that determines the basis chain. Critical points of $\widehat{f}$ that map to one critical point of f are transformed into one another by elements of π. Thus, the fact that the bases of the free $Z[\pi]$-modules C_λ are not uniquely chosen is due to the possibility that basis elements may be multiplied by elements of π.

When a function changes inside the same component of $\mathscr{F} - \mathscr{F}_1$, the number of its critical points (all of which are Morse) remains unchanged, and the family of separatrices of a gradient-like dynamical system is not rearranged. Hence the exact sequence $\{C_\lambda\}$ of free $Z[\pi]$-modules with fixed bases (which are determined almost uniquely in the above sense) is invariant.

Suppose that we have chosen an arbitrary function in general position on a manifold W. Associated to it, we have an exact sequence of free $Z[\pi]$-modules $\{C_\lambda\}$. According to the above construction, in order to determine whether there exists a function on W without critical points, i.e., whether W is diffeomorphic to the cylinder $\mathscr{M} \times I$, we have to describe the transformations of the sequence $\{C_\lambda\}$, when strata of codimension one intersect, that is, under bifurcations of general position in one-parameter families of functions. A necessary condition for the existence of a function without critical points is that $\{C_\lambda\}$ be reducible to the trivial sequence by a sequence of transformations α_k.

Singularities of codimension one are one of the following types:

(1) appearance of a critical point of type $\mathscr{A}_2$;
(2) a separatrix of a gradient-like vector field from a point $a_i^{(\lambda)}$ of index λ to another critical point $a_j^{(\lambda)}$ of the same index λ;
(3) simple tangency of separatrix disks along a separatrix going from a critical point of index $\lambda+1$ to a critical point of index λ in the direction of decreasing f (in this process, a pair of close-lying separatrices appear or disappear).

In the first case, tranversality of the intersection of the corresponding stratum means that the corresponding family is miniversal for the $\mathscr{A}_2$ singularities. In the second case, the transversality condition can be described as follows:

the value of f at one of the critical points of index λ (say, $f(a_i^{(\lambda)}) = z_i^{(\lambda)}$) is greater than that at the other ($f(a_j^{(\lambda)}) = z_j^{(\lambda)}$). This is true at a bifurcational value of the parameter, hence also at all sufficiently close ones. Choose a noncritical value z_0 of f between $z_i^{(\lambda)}$ and $z_j^{(\lambda)}$. Separatrices emanating from $a_i^{(\lambda)}$ and $a_j^{(\lambda)}$ (the first in the direction of decreasing f, the other in the direction of increasing f) cut out spheres of dimension $\lambda - 1$ and $n - \lambda$, respectively, on the n-dimensional level manifold $\{f = z_0\}$. In the general position, i.e., for a general value of the parameter, these spheres do not intersect. In the codimension-one situation (i.e., for a bifurcational value of the parameter) they may intersect transversally (see Figure 2, where $n = 3$ and $\lambda = 2$).

The appearance of a critical point of type $\mathscr{A}_2$ (in a generic way) implies the birth or death of a pair of critical points of consecutive indices λ and $\lambda + 1$. As one of these situations (birth) may be switched into the other (death) by changing the sign of the parameter, it will suffice to understand what happens to the exact sequence $\{C_\lambda\}$ in a situation, say, of birth. The birth of a pair of critical points results in the addition of free generators $\Delta^{(\lambda)}$ or $\Delta^{(\lambda+1)}$ to C_λ and $C_{\lambda+1}$, respectively, and the corresponding critical points are connected by a separatrix. Hence the basis element $\Delta^{(\lambda)}$ appears in the boundary $\partial \Delta^{(\lambda+1)}$ with coefficient 1 (provided that the orientation is suitably chosen). It can be shown that by changing the gradient-like vector field we can make the boundary $\partial \Delta^{(\lambda+1)}$ equal to $\Delta^{(\lambda)}$ and the boundary $\partial \Delta^{(\lambda)}$ zero (the latter follows from the former), at the same time ensuring that $\Delta^{(\lambda)}$ and $\Delta^{(\lambda+1)}$ do not occur in the expansions of the boundaries of other basis chains in terms of the basis elements. To prove this, we note that the families of separatrices emanating from a critical point of type $\mathscr{A}_2$ (in the direction of increasing or decreasing f), or those emanating from the pair of points created by that point, cut out on the manifolds lying below and above pairs of balls of certain dimensions (λ and $n - \lambda - 1$; see Figure 3, where $n = 2$, $\lambda = 0$). Since these intersections are homotopically trivial, there are no topological obstructions to the deformation of these balls in such a way that they do not intersect balls cut out by other families of separatrices. This can be achieved by a deformation of the gradient-like vector field. Of course, such a deformation may cause families of separatrices to bifurcate; nevertheless, modulo the transformations corresponding to such bifurcations (which will be described below), we may assume that separatrices emanating

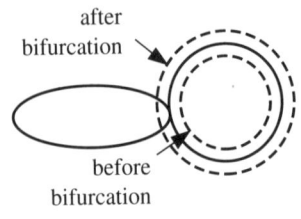

FIGURE 2

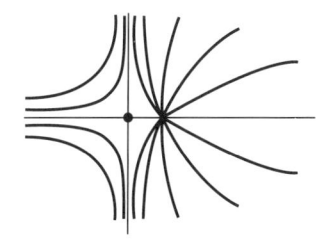

FIGURE 3

from points of type $\mathscr{A}_2$ do not "hit" other critical points of f. This is just the case in the situation described above. In other words, the appearance of a critical point of type $\mathscr{A}_2$ and the birth of a pair of nondegenerate critical points lead to the addition of a free term of the form $0 \leftarrow Z[\pi] \stackrel{id}{\leftarrow} Z[\pi] \leftarrow 0$ to the exact sequence $\{C_\lambda\}$, in the position corresponding to the modules $C_\lambda \leftarrow C_{\lambda+1}$.

Let us now find what happens to $\{C_\lambda\}$ under a bifurcation in which a separatrix goes from a critical point $a_i^{(\lambda)}$ of index λ to another critical point $a_j^{(\lambda)}$ of the same index ($Z_i^{(\lambda)} = f(a_i^{(\lambda)}) > f(a_j^{(\lambda)}) = Z_j^{(\lambda)}$). The corresponding change in the basis chains (whose number, of course, remains the same) can be described in detail in the general case. However, it will be easier to consider a representative low-dimensional case, e.g., $n+1 = 3$, $\lambda = 2$. The basis chain corresponding to $a_j^{(2)}$ is determined by the two-dimensional disk $\mathscr{D}_j^2$ swept out by the separatrices going from $a_j^{(2)}$ in the direction of decreasing f. The disk $\mathscr{D}_i^2$ of separatrices emanating from $a_i^{(2)}$ (which is also two-dimensional) does not contain a separatrix emanating from $a_j^{(2)}$ in the direction of increasing f. If the parameter value is close to the bifurcation parameter, $\mathscr{D}_i^2$ passes near such a separatrix and "glides" along $\mathscr{D}_j^{(2)}$ in one direction. At bifurcation it includes the separatrix but subsequently begins to "glide" along $\mathscr{D}_j^{(2)}$ in the other direction (see Figure 4 on p. 120). It follows that in the course of the bifurcation the chain $\Delta_j^{(2)}$ associated with $a_j^{(2)}$ remains unchanged, but the chain $\Delta_i^{(2)}$ associated with $a_i^{(2)}$ is modified by the addition of $\Delta_j^{(2)}$ with a coefficient depending on the choice of chain orientation and on points of the orbits $\pi a_i^{(2)}$ and $\pi a_j^{(2)}$ that were chosen to define the basis elements in the $Z[\pi]$-module C_2. The other basis chains remain, of course, unchanged. Thus, the bifurcation modifies one basis element (the ith) by adding a multiple of some other basis element (the jth).

One immediately sees that a bifurcation of the third type (i.e., tangency of separatrix disks along a separatrix connecting critical points of indices $\lambda+1$ and λ) leaves the sequence $\{C_\lambda\}$ unchanged. This is easily understood if one observes that before the bifurcation the points are not connected by a separatrix, while after the bifurcation these appear as two close separatrices with

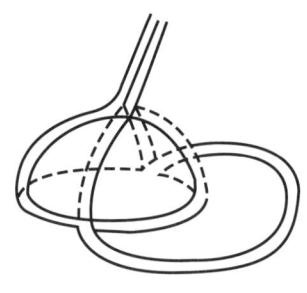

FIGURE 4

multiplicities $+1$ and -1, whose contributions to the incidence coefficient cancel out.

Therefore, to an intersection of a codimension-one stratum there corresponds either a stabilization of the exact sequence $\{C_\lambda\}$ (addition or removal of a free term $0 \leftarrow Z[\pi] \stackrel{\mathrm{id}}{\leftarrow} Z[\pi] \leftarrow 0$) or a change of the basis $\alpha_{ij}(\pm g)$ in one of the free $Z[\pi]$-modules C_λ as was described above. Therefore, the equivalence class of an exact sequence $\{C_\lambda\}$ modulo the equivalence relation defined by these transformations is preserved under deformation of functions. As mentioned earlier, the set of equivalence classes of exact sequences of free $Z[\pi]$-modules $\{C_\lambda\}$ with bases forms the group $K_1(Z[\pi])$, or the Whitehead group $\mathrm{Wh}(\pi)$, if the nonuniqueness of the bases due to the possibility of multiplying each basis element by elements of $\pm\pi$ is taken into account. Therefore, with any h-cobordism $(W, \mathscr{M})$, we have associated an element $\tau(W, \mathscr{M})$ of $\mathrm{Wh}(\pi)$, known as the *torsion*, and proved that if $\tau(W, \mathscr{M}) \neq 0$ in $\mathrm{Wh}(\pi)$, then there are no functions without critical points on the h-cobordism $(W, \mathscr{M})$, i.e., the latter is not diffeomorphic to the cylinder $(\mathscr{M} \times I, \mathscr{M} \times 0)$.

Of course, this argument does not allow us to assert that if $\tau(W, \mathscr{M}) = 0$, then the h-cobordism $(W, \mathscr{M})$ is diffeomorphic to a cylinder, or that there exists an h-cobordism with nontrivial torsion (and therefore not diffeomorphic to a cylinder). In fact, that is the content of the h-Cobordism Theorem (see [2]).

THEOREM. *For $n = \dim \mathscr{M} \geq 5$, the h-cobordism $(W, \mathscr{M})$ is diffeomorphic to a cylinder if and only if $\tau(W, \mathscr{M}) = 0$ in $\mathrm{Wh}(\pi)$ $(\pi = \pi_1(\mathscr{M}))$. Furthermore, for any element $\tau \in \mathrm{Wh}(\pi)$ there exists an h-cobordism $(W, \mathscr{M})$ with torsion τ; the torsion establishes a correspondence between the h-cobordisms $(W, \mathscr{M})$ with a fixed base $\mathscr{M}$ and the group $\mathrm{Wh}(\pi)$, which is one-to-one up to diffeomorphisms of the former.*

Now we discuss how to compute the set (in this particular case, the group) $\pi_0(P_i(\mathscr{M}))$ of components of the semigroup $P_i(\mathscr{M})$ of pseudo-isotopies of a manifold $\mathscr{M}$. As we have seen, $\pi_0(P_i(\mathscr{M})) \cong \pi_1 \ (\mathscr{F} < \mathscr{F}(0))$, where $\mathscr{F}$ is the space of C^∞ functions $(\mathscr{M} \times I, \mathscr{M} \times 0, \mathscr{M} \times 1) \to (I, 0, 1)$ and $\mathscr{F}(0)$ the subspace of functions without critical points. The construction described above yields a mapping of $\pi_1(\mathscr{F}, \mathscr{F}(0))$ into the group of equivalence classes of products of formal generators $\alpha_{ij}(\pm g)$ that become identity matrices when the elements $\alpha_{ij}(\pm g)$ are replaced by appropriate matrices; the equivalence relation is defined by paths around codimension-two strata. Therefore, we must determine the relations between the elements $\alpha_{ij}(\pm g)$ corresponding to codimension-two singularities.

Codimension-two singularities may be of the following types:

(1) appearance of separatrix connections between two different pairs of points of the same indices;

(2) a separatrix emanating from a point $a_i^{(\lambda)}$ of index λ "hitting" another point $a_j^{(\lambda)}$ of index λ, and a separatrix emanating from the latter "hitting" a third point $a_k^{(\lambda)}$ of the same index;

(3) a separatrix going from a point of index λ in the direction of decreasing f to a point of index $\lambda + 1$;

(4) simple tangency of disks of separatrices along a separatrix connecting two points of index λ;

(5) second-order tangency of disks of separatrices along a separatrix from a point of index λ to a point of index $\lambda - 1$ in the direction of decreasing f;

(6) appearance of a critical point of a function of type $\mathscr{A}_3$.

The versal deformations of these singularities and the corresponding bifurcation diagrams that determine the products of elements $\alpha_{ij}(\pm g)$ may be described as follows.

In the first case, the two separatrix connections deform independently, and so a versal deformation is the direct product of two (one-parameter) versal deformations, one for each of these connections. In the bifurcation diagram shown in Figure 5, the indices indicate the points joined by separatrix connections. It is easy to see that the corresponding product of elements $\alpha_{ij}(\pm g)$ is equal to the commutator

$$[\alpha_{ij}(g_1)\alpha_{k\ell}(g_2)] = \alpha_{ij}(g_1), \alpha_{k\ell}(g_2)(\alpha_{ij}(g_1))^{-1}(\alpha_{k\ell}(g_2))^{-1},$$

and the corresponding relation has the form $[\alpha_{ij}(g_1), \alpha_{k\ell}(g_2)] = 1$ (the pairs (i, j) and (k, ℓ) have no common elements).

In the second case (separatrix connections between the pairs of points $(a_i^{(\lambda)}, a_i^{(\lambda)})$ and $(a_j^{(\lambda)}, a_k^{(\lambda)})$), there is a separatrix connection between $a_i^{(\lambda)}$

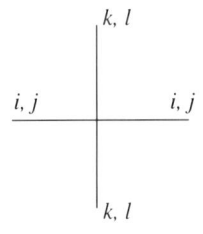

FIGURE 5

and $a_k^{(\lambda)}$ on a codimension-one stratum adjacent to the codimension-two stratum in question (see Figure 6, where $n + 1 = 2$, $\lambda = 1$; the figure shows the integral curves of a gradient-like dynamical system for a function on codimension-two stratum (in the center), on the stratum corresponding to the separatrix connection between $a_i^{(\lambda)}$ and $a_k^{(\lambda)}$ (left), and on its continuation behind the codimension-two stratum (right)). The corresponding bifurcation diagram is shown in Figure 7. The indices of the points linked by a separatrix connection are indicated near each codimension-one stratum. The equivalence relation (determined by a product of elements $\alpha_{ij}(\pm g)$) is

$$[\alpha_{ij}(g_1), \alpha_{jk}(g_2)] = \alpha_{ik}(g_1 g_2).$$

It is easy to verify that bifurcations of the third and fourth types do not produce nontrivial relations. Bifurcations of the fifth type give the relation $\alpha_{ij}(g)\alpha_{ij}(-g) = 1$.

It is somewhat more difficult to understand the relation corresponding to a bifurcation of the sixth type (birth of a critical point of type $\mathscr{A}_3$). There is no difficulty in drawing the bifurcation diagram, which is just an ordinary semicubical parabola, on whose branches the functions have critical points of type $\mathscr{A}_2$. The difficulty is that the pairs of basis chains associated with singularities of type $\mathscr{A}_2$ do not split off as direct summands from the respective exact sequences of free $Z[\pi]$-modules. The reason is that the separatrices of the gradient vector field (defined in the most natural way) connect critical points of type $\mathscr{A}_2$ with nondegenerate critical points which have split off from a singularity of type $\mathscr{A}_3$. To achieve splitting, we have to modify the gradient-like vector field, and this may result in new codimension-one bifurcations (separatrix connections). The detailed analysis of this case shows that

FIGURE 6

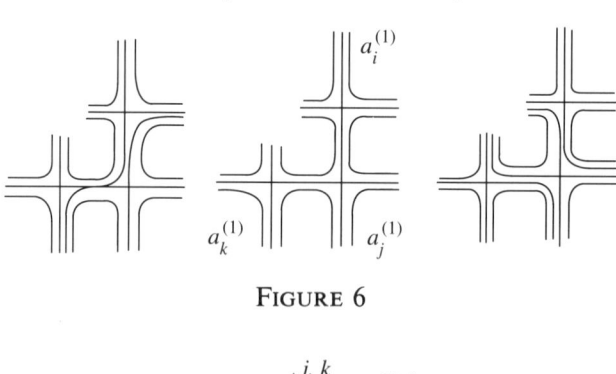

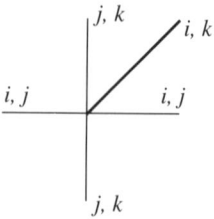

FIGURE 7

under the corresponding relation a product $\alpha_{ij}(g)\alpha_{ij}(-g^{-1})\alpha_{ij}(g)$ is equivalent to transposing the ith and jth basis elements and multiplying them by g and $-g^{-1}$, respectively (i.e., in the ring of matrices this is the identity $\alpha_{ij}(g)\alpha_{ij}(-g^{-1})\alpha_{ij}(g) = \begin{pmatrix} 0 & g \\ -g^{-1} & 0 \end{pmatrix}$, where only the block corresponding to the ith and jth basis elements is shown).

The above construction associates to each element of $\pi_1(\mathscr{F}, \mathscr{F}(0))$, i.e., to any path in $\mathscr{F}$ starting and ending in $\mathscr{F}(0)$, a product of elements $\alpha_{ij}(g)$ which, when $\alpha_{ij}(g)$'s are replaced by the appropriate matrices, becomes the identity matrix modulo the above relations. One immediately sees that these relations are exactly the same as those defining the group $\text{Wh}_2(\pi)$ (and, moreover, the relations corresponding only to a bifurcation of separatrix connections are those defining $K_2(Z[\pi])$). We have thus defined a homomorphism

$$\pi_0(P_i(\mathscr{M})) \cong \pi_1(\mathscr{F}, \mathscr{F}(0)) \to \text{Wh}_2(\pi).$$

THEOREM (Volodin, Hatcher, and Wagoner). *For* $\dim \mathscr{M} = n \geq 6$, $\pi_0(P_i(\mathscr{M}))$ *is the direct sum of the second Whitehead group* $\text{Wh}_2(\pi)$ *and the group* $\text{Wh}_1(\pi, Z_2 \times \pi_2)$ *defined functionally by the fundamental group* $\pi = \pi_1(\mathscr{M})$ *and the second homotopy group* $\pi_2 = \pi_2(\mathscr{M})$ *with the action of* π.

References

1. I. A. Volodin, *Generalized Whitehead groups and pseudo-isotopies*, Uspekhi Mat. Nauk **27** (1972), no. 5, 229–230. (Russian)
2. J. Milnor, *Whitehead torsion*, Bull. Amer. Math. Soc. **72** (1966), 358–426.
3. _____, *Introduction to algebraic K-theory*, Princeton Univ. Press, Princeton, N. J., 1971.
4. A. E. Hatcher, *The second obstuction for pseudo-isotopies*, Bull. Amer. Math. Soc. **78(6)** (1972), 1005–1008.
5. J. B. Wagoner, *Algebraic invariants for pseudo-isotopies*, Proceedings Liverpool Singularities Symposium. II, Lecture Notes in Math., vol. 209, Springer-Verlag, Berlin, 1971, pp. 164–190.

Translated by LEONID HANIN

Singularities in Optimization Problems

A. A. DAVYDOV

A usual phenomenon in the history of singularity theory is that a typical singularity, once discovered in the context of a certain problem, will generally reappear frequently in other problems. Problems of variational calculus and optimal control have turned out to be rich in both already-known and new typical singularities (see, e.g., [1, 4, 11, 12, 14]). Here we will make the acquaintance of certain singularities encountered in the study of a two-dimensional control system in general position.

To be precise, we assume that the phase space of the control system is a smooth two-dimensional manifold M without boundary, and that the set of values of the control parameter is a disjoint union of finitely many compact smooth manifolds and contains at least two distinct points. A control system is given by a mapping F of a bundle P over M with fiber V into the space of the tangent bundle TM, such that the diagram

$$\begin{array}{ccc} P & \xrightarrow{F} & TM \\ & \searrow_{\tau} & \downarrow \pi \\ & & M \end{array}$$

is commutative, where τ and π are the projections of the corresponding bundles. We identify the space of control systems with the set of such mappings and endow it with Whitney's fine C^4-topology. A system in general position is an element of an open dense subset of this space in this topology.

The indicatrix at a point x of the phase space is defined to be the set $F(\tau^{-1}(x))$ of admissible velocities at x. The control problem is to reach a specified target moving at all times at an admissible velocity, which is a piecewise continuous function of time. Control targets may differ; we will

---

1991 *Mathematics Subject Classification.* Primary 58C27, 57R45; Secondary 34K35, 58A20, 34C05, 49A10, 49E99.

discuss the following three, in various degrees of elaboration:

A: to conserve the initial state of the system or permit it to drift in an arbitrarily small neighborhood of the initial state;

B: to let the system drift in an open region of states while maintaining its ability to return to any previous state, beginning with the initial state;

C: to steer the system from a given set of initial states (starting set) to a fixed system state in (C_1) an arbitrary time; (C_2) a fixed time; (C_3) the least possible time.

The following example illustrates the fact that in each case the control target cannot always be reached from an arbitrary point of phase space of the system.

Swimmer in a current. On the plane $R^2_{x,y}$ a current with vector field $(-x, -\beta y)$ carries a swimmer to the origin. The swimmer himself, in still water, can swim in any direction with the unit speed. Here the vectors $(-x + \cos u, -\beta y + \sin u)$, $0 \leq u < 2\pi$, form the indicatrix of admissible velocities at a point (x, y) of the plane.

Problem A is solvable in the region $x^2 + \beta^2 y^2 \leq 1$. The convex hull of the indicatrix of admissible velocities at each point of the region contains the zero velocity. This region is the closure of the swimmer's *local transitivity zone* (i.e., the set of all points x such that for any x and any point y sufficiently close to x, either of x or y is attainable from the other in a small time).

The inequality $x^2 + \beta^2 y^2 > 1$ defines the *steep domain* of the control system, that is, the set of all points at which the convex hull of the indicatrix does not contain the zero velocity. Problem A is not solvable in the steep domain, since the swimmer, whatever the direction $(\cos u, \sin u)$ in which he tries to swim, will not be able to oppose the current.

It is clear that problem A is always solvable in the local transitivity zone of the system and unsolvable in the steep domain. However, the sets of states for which problem A is solvable or unsolvable do not necessarily coincide with the local transitivity zone and steep domain, respectively. The reason is that in general position the boundary between the local transitivity zone and the steep domain is a smooth curve with singularities. Only a few points of the curve do not belong to the local transitivity zone. For some of these points problem A is solvable; for others it is not.

Problem B is solvable if and only if the initial state of the system belongs to a *nonlocal transitivity zone* of the system (a nonlocal transitivity zone is an open domain in the phase space, defined as the intersection of the positive and negative orbits of any of its points. A control system may have several nonlocal transitivity zones). For a better understanding of the structure of the solvability region of problem B in our example, we conduct a preliminary investigation.

At any point of the steep domain the directions of the admissible velocities form an angle less than $180°$, whose sides are called the *limit directions at the point*. Thus, a two-valued field of limit directions is defined in the steep domain. The integral curves of this field are called *limit curves*.

On the boundary of the steep domain the field of limit directions defines a field of straight lines. In general position, this field of straight lines may have first-order contact with the boundary at a nonsingular point. At such points the field of limit directions has typical singularities. Before describing them, we note that, in our example, the limit curves are just the phase curves of a first-order differential equation in which the derivative appears implicitly:

$$(x\,dy - \beta y\,dx)^2(x^2 + \beta^2 y^2 - 1) = (x\,dx + \beta y\,dy)^2,$$

and the field of limit directions touches the boundary of the steep domain at four points $(\pm 1, 0), (0, \pm 1/\beta)$.

The equation $F(x, y, p) = 0$ in general position defines a smooth surface in the space of 1-jets of functions $y(x)$ (with coordinates $x, y, p = dy/dx$). The restriction of the projection $(x, y, p) \to (x, y)$ to this surface is called the *folding of the equation*. The critical points of the folding of an equation form the criminant of the equation, and the critical values describe the discriminant curve. The field of contact planes $dy = p\,dx$ cuts out on the surface of the equation a field of directions, the images of whose integral curves, when folded, are mapped onto the phase curves of the equation (note that on the criminant this field of directions touches the kernel of the folding). In general position this field of planes and the criminant of the equation may have first-order contact at various points. At each of these points the folding of the equation has a critical point of Whitney's fold type and the field of directions of the equation has a nondegenerate singular point that is either a saddle, a node, or a focus. The image of such a singular point under the folding of the equation is called a folded saddle, node, or focal singular point of the equation $F(x, y, p) = 0$, respectively. The germ of the equation in general position at a folded singular point is mapped by a C^∞ diffeomorphism of the plane into the germ at zero of the equation

$$y = (dy/dx + kx)^2$$

where $k < 0, 0 < k < \frac{1}{8}$, and $\frac{1}{8} < k$ for a saddle, node, and focus, respectively (applying a suitable homeomorphism, one can make $k = -1, k = 0$, and $k = 1$, respectively). A detailed exposition of these results for implicit differential equations may be found in [2, 10].

In our example, the discriminant curve is simply the boundary of the steep domain; the criminant touches the contact plane at a point corresponding to a point of contact of the field of limit directions with the boundary. Thus, in general position, the pattern of the limit curves in a small neighborhood of such a point is exactly the same as that of the phase curves in a small neighborhood of one of the singular points—a folded saddle, node, or focus of an implicit first-order differential equation. It is easy to verify that at the points $(\pm 1, 0)$ we have folded saddles, and at $(0, \pm 1/\beta)$ folded nodes. Each saddle point is connected to each node by a separatrix, so that there are exactly four connecting separatrices. The swimmer's nonlocal transitivity zone is the open domain bounded by the curve consisting of these four folded singular points and the four separatrices connecting them. It is precisely in this domain that the swimmer's problem B is solvable.

Clearly, in the case of an arbitrary starting set, problems C_1, C_2, C_3 are not solvable for sufficiently remote points of the plane. For example, if the starting set lies in a domain $x^2 + \beta^2 y^2 \leq a$, $a > 1$, no point outside it is attainable from the starting set. The swimmer cannot oppose the current on the boundary of the domain, but is carried into the interior of the domain.

It may happen that problems C_2, C_3 are unsolvable even if problem C_1 is solvable. It may even happen that the system can be steered to a fixed state, but this state cannot be reached in a fixed or even minimal time. For example, in the control system defined in the plane by the two fields of admissible velocities $(1, \pm 1) \cdot 1/(1 + y^2)$, the state $(1, 0)$ is attainable from the state $(0, 0)$ in any time $t \in (1, \frac{13}{12})$, but not attainable in time $t \leq 1$ or $t > \frac{13}{12}$.

Stability of the steep domain and of its singularities. A more complete exposition of results concerning the steep domain and singularities of fields of limit directions of systems in general position may be found in [11].

THEOREM 1. *The steep domains of a control system in general position and any system sufficiently close to it may be mapped into each other by a C^∞-diffeomorphism of the phase space that is close to the identity.*

Thus, the steep domain of a control system in general position and the singularities of its boundary are stable under small perturbations of the system. In order to describe these singularities better, we define some new notions.

A boundary point of the steep domain is called a *zero-passing* point if there exists an isolated value of the control parameter with zero velocity at this point. Note that there are no zero-passing points if there are no isolated values of the control parameter.

A value of the control parameter is called a *limit value* at a point of the phase space if the velocity it defines at this point belongs to a limit direction there.

A boundary point of the steep domain is called a *zero-point* if there is a unique limit control there; it is called a *nonzero-passing* point if the number of different limit controls there is exactly two; one of them gives the system nonzero velocity at the point, while the other gives it zero velocity and is a nonisolated value of the control parameter.

EXAMPLE. $M = R^2_{x,y}$, $V = S^1_u \cup \{10\}$, $0 \leq u < 2\pi$; the velocity indicatrix at a point (x, y) is the set of vectors $(x/\sqrt{1+x^2} + \cos u, (1+y)/\sqrt{1+x^2} + \sin u)$, plus the vector $(-2x, 2-y)$ corresponding to the value 10 of the control parameter. For this system, $(\pm 1, 0)$ are nonzero-passing points, the points of the interval $(-1, 1)$ of the x-axis are zero-points, $(0, 2)$ is a zero-passing point.

A boundary point of the steep domain will be called a ∂-*passing point* or a *double ∂-passing point* if the velocity indicatrix at the point does not contain the zero velocity, and the number of different limit controls there is two or three, respectively.

THEOREM 2. *For a control system in general position*:

(I) *the germ of the steep domain at each point z of its boundary is C^∞-diffeomorphic to the germ at zero of one of the following seven sets*:
1) $\{y \neq 0\} \cup \{x < 0\}$, 2) $y \neq 0$, 3) $|y| > |x|$, 4) $y > |x|$, 5) $y < |x|$, 6) $y > 0$, 7) $\{x < \sqrt{y}\} \cap \{y > 0\}$;

(II) *each of these singularities occurs only when the number of different values of the control parameter is, respectively,* 1) 2; 2) 2; 3) 3; 4) > 3; 5) > 2; 6) > 2; 7) ∞;

(III) *the point z is, respectively,* 1) *a zero-passing point*; 2) *a ∂-passing point*; 3) *a double ∂-passing point*; 4) *a double ∂-passing point*; 5) *a zero-passing point*; 6) *either a ∂-passing point or a zero-point*; 7) *a nonzero-passing point.*

Realizations of these singularities are easy to construct, so we shall not present them here.

The complement of the steep domain of a control system in general position is the closure of its local transitivity zone. Therefore, Theorem 2 also describes the singularities of the closure. In the entire closure, only a few points of the boundary of the steep domain do not belong to the local transitivity zone itself. Each of these points is either a zero-passing point or a point of contact of the field of limit directions and the boundary.

Stability of the attainable set and its singularities. Let us fix the starting set, that is, the set of points from which we are permitted to start a motion. The set $\mathscr{D}(t)$ of all points of the phase space, each being attainable from at least one point of the starting set in time t, called the *attainable set at time t*. The union of the sets $\mathscr{D}(t)$ over all $t \geq 0$ is called the *attainable set $\mathscr{D}$*. A state of the system, that is, of a point z of the phase space, belongs to $\mathscr{D}$ (to $\mathscr{D}(t)$) if and only if there is at least one point of the starting set from which the system can be steered by an admissible control to z (exactly in time t). The attainable sets $\mathscr{D}$ and $\mathscr{D}(t)$ are important characteristics of a control system and are being actively studied from various points of view (see, e.g., [5, 8, 16]).

In general, these sets may have singularities on their boundaries, even when neither the field of indicatrices nor the starting set have singularities.

For example, if the swimmer starts at the origin, the attainable set is just the nonlocal transitivity zone, whose boundary has singularities at the points $(0, \pm\frac{1}{\beta})$ if β is not an integer.

A point on the boundary of the attainable set will be called a singular point of type ρ, $1 \leq \rho \leq 6$, if the germ of the closure of the attainable set at this point is C^∞-diffeomorphic to the germ at zero of one of the following sets:

- the set $y \underset{(<)}{\geq} |x|x^{\rho-1}$ if $\rho = 1, 2$;
- the set $y \geq |x|x^2$ if $\rho = 3$;
- the set $\varepsilon x^\alpha \leq y \leq x$, where $\varepsilon = 1$, if $\rho = 4$;
- the set
$$y \leq \begin{cases} C_1|x|^\alpha, & \text{if } x \leq 0, \\ C_2 x^\alpha, & \text{if } x > 0, \end{cases}$$
where $C_1, C_2 \in R$, $C_1 C_2 \neq 0$, if $\rho = 5$;
- the set $y \leq h(x)$, if $\rho = 6$, where the germ at zero of the boundary of this set is the germ at zero of the closure of the union of two nonsingular phase curves of the folded node
$$y = (dy/dx + \alpha(\alpha+1)^{-2} x/2)^2$$
which enter zero from opposite directions.

In each case, $\alpha > 1$ is not an integer.

Assume, in addition, that the phase space is compact, and that the starting set is a circle smoothly imbedded in the phase space.

THEOREM 3. *For a control system in general position*
(1) *the closure of its attainable set is the closure of its own interior;*
(2) *the boundary of the attainable set is either empty or a smoothly imbedded curve with singular points of type ρ, $1 \leq \rho \leq 6$.*

THEOREM 4. *The attainable sets of a control system in general position and of any system sufficiently close to it are mapped onto each other by a homeomorphism of the phase space that is close to the identity; the homeomorphism can be chosen so that it is a C^∞-diffeomorphism everywhere except possibly at singular points of type 4, 5, and 6 on the attainable boundary.*

Note that in the general position there are no points of types 4 and 5 if the control parameter has no isolated values. The reason is that singular points of these types are always zero-passing points. In contrast, if the set of values of the control parameter is zero-dimensional, singular points of type 6 do not occur, since singular points of this type are always zero points. At each of these points the family of limit curves has a folded node.

Singularities of type 1, 2, 3 will appear for all possible dimensions of the manifolds occurring in the set of values of the control parameter.

There is a type 6 singularity at the point $(0, \pm\frac{1}{\beta})$ on the boundary of our swimmer's nonlocal transitivity zone, if β is not an integer. We now illustrate singularities of types 1 to 5. To this end, consider the following example.

Ship in current of water and wind. In a plane sea $R^2_{x,y}$ an inertialess ship has gone adrift and is carried in turn either by a current of water with field v or by wind above the water with field w.

If $v = (-x + 1; -2.1y + 2.1)$ and $w = (-x - 1; -2.1y - 2.1)$, then the current field carries the ship to the point $(1, 1)$, while the wind field carries it to $(-1, -1)$ (each of these points is a nondegenerate stable node of the corresponding current). Take the circle $(x + 3)^2 + y^2 = 4$ as the starting set. The attainable set (the search area for the ship) is bounded by the union of the following four lines: the part of the phase curve of the water field from the point A at which the field touches the upper half of the circle $y = \sqrt{4 - (x+3)^2}$ to the point $(1, 1)$; the part of the phase curve of the wind field from the point B at which the field touches the lower half of the circle $y = -\sqrt{4 - (x+3)^2}$ to the point $(-1, -1)$; the left arc AB of the latter circle; and the part of the phase curve of the wind field from $(1, 1)$ to $(-1, -1)$. The boundary of the attainable set contains singularities of type 2 at points A and B, of type 4 at $(1, 1)$, and of type 5 at $(-1, -1)$.

Let now the current field be $v = (1, -1)$, the wind field $w = (-1, -1)$, and take the curve $x = \cos t$, $y = \cos(2t - 0.2)$, $0 \le t \le 2\pi$, as the starting set. At the point of intersection of the phase curve of the current field that touches the starting set in the second quadrant and the phase curve of the wind field that touches it in the first quadrant, there is a type 1 singularity of the attainable set, of the form $y \le |x|$. (Note that the starting set is a curve with a point of self-intersection, but this point lies inside the attainable set and is thus nonessential.) A type 1 singularity of the form $y \ge |x|$ is obtained at zero in the system $v = (-1, 1)$, $w = (1, 1)$, and also in the field of admissible speeds $(-2x, -y)$ with the starting set $x^2 + (y - 2)^2 = 1$.

Finally, a type 3 singularity at zero occurs in the system $v = (-1, 0)$, $w = (1, x^2 - y)$ with the initial set lying in the region $y > x^2$. Here the ship's search area lies over the union of the parts of the phase curves of the wind field and of the current field that emanate from zero.

The following phenomenon is related to type 3 singularities. In general position two smooth vector fields in the plane are collinear on some curve. They define a field of straight lines on this curve, which may touch it at certain points with first-order contact. It turns out that the germ of the family of limit lines of the system defined by these two fields at such a point of contact is C^∞-diffeomorphic to the germ at zero of the family of limit curves of the system defined by either the field $v = (-1, 0)$, $w = (1, x^2 - y)$, or $v = (1, 0)$, $w = (1, x^2 - y)$. Note that in the analytic case the normal form of the germ at such a point of contact of the family of limit curves contains function moduli related to the moduli described by Écalle [18], and Voronin [9].

It is easy to verify that the above realizations are stable under small perturbations of the system. A system close to the original system has singularities of the same types at nearby points.

Singularities of minimum functions over the inverse image. Consider the function on the attainable set defined at each point as the infimum of the times at which the point is reachable from the starting set. This function is nonnegative on the starting set. In general position it generally has singularities, and the singularities on a level t curve of this function will yield singularities on the boundary of the attainable set in time t. In general position, for any bounded domain of the phase space, the list of germs of these singularities, up to R^+-equivalence (that is, up to a C^∞-diffeomorphism of phase space and the addition of a smooth function) is exhausted by the germs at zero of the following four functions:

$$1)\; 0; \quad 2)\; -|x|; \quad 3)\; -(||x|+y|+|x|); \quad 4)\; \min\{w^4+w^2y+wx \mid w \in \mathbb{R}\}, \tag{1}$$

if

a) the starting set is a smoothly imbedded circle,
b) the velocity indicatrices are strictly convex, and
c) each of them contains the zero vector inside its convex hull.

Note that R^+-equivalence obscures a number of essential details. For example, if a perturbation in a plane homogenous medium originates in two ellipses $(x\pm 2)^2 + 2y^2 = 1$, R^+-equivalence does not notice that singularities of the response time at $(0, 0)$ and $(0, 1)$ are even topologically distinct.

There is no list of typical singularities for the response time if one of conditions b), c) does not hold. If condition c) alone fails to hold, the list is probably contained in the list of typical singularities of the minimum function over the inverse image of a two-dimensional image. This function is defined as follows: let N be a smooth manifold without boundary; consider a smooth function $f : N \to \mathbb{R}$ and a smooth proper mapping $g : N \to \mathbb{R}^m$ with $n = \dim N \geq m$. The minimum function over the inverse image for the couple (f, g) is defined as follows:

$$G(q) = \min\{f(p) \mid p \in g^{-1}(q), q \in g(N)\}.$$

The function G need not be smooth, or even continuous. In general position it has the standard singularities, if $m = 1, 2$; the finite list of singularities is given below. The singularities of this function for $m > 2$ have been thoroughly investigated only in the case where g is a fibration [6, 7, 14]; in particular, for $m = 2$, this case yields just the four typical singularities listed above. For $n > m = 2$, typical singularities of the minimum function over the inverse image are realized in two-dimensional control systems as singularities of the response time which are stable under small perturbations of these systems. There seem to be no obstacles to prevent us from obtaining similar realizations of the singularities of this function for $n > m > 2$.

Germs of two functions are Γ-*equivalent* if there exists a C^∞-diffeomorphism of the spaces of their graphs that is fibered over the spaces in which they are defined and carries the germs into each other. A pair in general position is a pair (f, g) in an open, dense subset of the space of pairs in Whitney's fine C^4-topology.

THEOREM 5. *If $m = 1$ and $n \geq 1$ and if the pair is in general position, the germ at each point of the image Γ of the minimum function over the inverse image is equivalent to the germ at zero of one of the following four functions*:

$$1)\ 0; \quad 2)\ -|x|; \quad 3)\ -\sqrt{x}; \quad 4)\ \min\{-\sqrt{x}; 1\}. \qquad (2)$$

THEOREM 6. *If $m = 2$ and $n \geq 2$ and if a pair is in general position, the germ at each point of the image Γ of the minimum function over the inverse image is equivalent to the germ at zero of one of the following thirteen functions*:

1) 0; 2) $-|x|$; 3) $-||x|+y|-|x|$;

4) $\min\{w^4 + w^2 x + wy | w \in R\}$; 5) $-\sqrt{x}$; 6) $\min\{-\sqrt{x}, 1\}$;

7) $\min\{-\sqrt{x}, -|y|+1\}$; 8) $\min\{-\sqrt{x}, -\sqrt{y}+1\}$;

9) $\min\{-\sqrt{x}, -\sqrt{y}+1, 2\}$; 10) $\min\{-\sqrt{x}, y\}$;

11) $-\sqrt{x}|y|$; 12) $\min\{-\sqrt{x}|y|, 1\}$; 13) $\min\{w \in R | w^3 + wx + y = 0\}$.

Singularities 11), 12) *do not occur for $n > 2$; singularity* 4) *does not occur for $n = 2$*.

Endow both $C^\infty(N, \mathbb{R})$ and the space of proper mappings $C_p^\infty(M, R)$ with Whitney's fine C^4-topology. A generic mapping is a member of an open dense subset of the space of such mappings in the given topology.

THEOREM 7. *Let $m = 1, 2$, $n \geq m$; let g (resp., f) be a fixed generic mapping and f (resp., g) a generic mapping. Then the germ at each point of the image of g of the minimum function over the inverse image is Γ-equivalent to the germ at zero of one of the functions in the list of singularities for $m = 1$ and $m = 2$, respectively.*

A fixed mapping g (resp., f) in general position is a mapping that has the Morin (resp., Morse) critical points [3, 15].

We now illustrate a type 13 singularity from the list in Theorem 6 as a singularity of the response time (realizations of type 1–10 singularities are simpler to obtain, while type 11, 12 singularities do not occur for $n > 2$). The velocity indicatrix at a point (x, y) is formed by the velocities $(\cos u, x + \sin u)$, $0 \leq u < 2\pi$. The response time has a type 13 singularity at $(1, 0)$, if the starting set is the circle $(x - a)^2 + y^2 = b^2$, which touches the limit curve entering the point $(1, 0)$ of the boundary of the steep domain (the equation of this curve is $y = (\ln(x + \sqrt{x^2 - 1}) - x\sqrt{x^2 - 1})/2$). This can be verified using the Pontryagin maximum principle [17].

Using the same principle, one can show that the minimum attainability time in a small neighborhood of the point $(1, 0)$ is given by orbits of the system that emanate at time $t = 0$ from points $x = a + b\cos\phi$, $y = b\sin\phi$, where $b > 0$, $\pi < \phi < 3\pi/2$, in the starting set and correspond to a continuous control $u(t)$ such that $u(0) = \phi$, $\cot u(t) = -t + \cot\phi$. These orbits sweep a smooth surface in x, y, t space. The restriction of the projection $(x, y, t) \to (x, y)$ to this surface has a critical point at a point $(1, 0, t_0)$ of this surface, of the Whitney fold type.

References

1. V. I. Arnol'd, *Singularities in variational calculus*, Itogi Nauki i Tekhniki. Sovremennye Problemy Matematiki. Noveĭshie Dostizheniya, vol. 22, VINITI, Moscow, 1983, pp. 3–55; English transl. in J. Soviet Math. **27** (1984).
2. V. I. Arnol'd and Yu. S. Il'yashenko, *Ordinary differential equations*, Itogi Nauki i Tekhniki. Sovremennye Problemy Matematiki. Fundamental'nye Napravleniya, vol. 1, VINITI, Moscow, 1985, pp. 5–150; English transl. in Encyclopedia Math. Sci., vol. 1, Springer-Verlag, Berlin, 1988.
3. V. I. Arnol'd, A. N. Varchenko, and S. M. Guseĭn-Zade, *Singularities of differentiable maps.* I, "Nauka", Moscow, 1982; English transl., Birkhäuser, Bäsel, 1985.
4. V. I. Arnol'd, *The theory of catastrophes*, "Znanie", Moscow, 1985. (Russian)
5. M. M. Baitman, *On domains of controllability on the plane*, Differentsial'nye Uravnenya **14** (1978), 579–593; English transl. in Differential Equations **14** (1978).
6. L. N. Bryzgalova, *Singularities of the maximum of functions depending on parameters*, Funktsional. Anal. i Prilozhen. **11** (1977), no. 1, 59–60; English transl. in Functional Anal. Appl. **11** (1977).
7. _____, *On the maximum functions of a family of functions depending on parameters*, Funktsional. Anal. i Prilozhen. **12** (1978), no. 1, 66–67; English transl. in Functional Anal. Appl. **12** (1978).
8. A. M. Vershik and V. Ya. Gershkovich, *Geometry of a nonholonomic sphere of three-dimensional Lie groups*, Geometry and Singularity Theory in Nonlinear Equations, Izd. Voronezh. Univ., Voronezh, 1987, pp. 61–75. (Russian)
9. S. M. Voronin, *An analytical classification of pairs of involutions and its applications*, Funktsional. Anal. i Prilozhen. **16** (1982), no. 2, 21–29; English transl. in Functional Anal. Appl. **16** (1982).
10. A. A. Davydov, *The normal form of a differential equation which is not solvable for the derivative, in the neighborhood of a singular point*, Funktsional. Anal. i Prilozhen. **19** (1985), no. 2, 1–10 (Russian); English transl. in Functional Anal. Appl. **19** (1985).
11. _____, *Singularities of fields of limit direction of two-dimensional control systems*, Mat. Sb. **136** (1988), no. 3, 477–499; English transl. in Math. USSR-Sb. **64** (1989), no. 2.
12. V. M. Zakalyukin, *Reconstructions of wave fronts depending on one parameter*, Funktsional. Anal. i Prilozhen. **10** (1976), no. 2, 69–70; English transl. in Functional. Anal. Appl. **10** (1976).
13. _____, *Singularities of convex hulls of smooth manifolds*, Funktsional. Anal. i Prilozhen. **11** (1977), no. 2, 76–77; English transl. in Functional Anal. Appl. **11** (1977).
14. V. I. Matov, *Topological classification of germs of minimum and minimax functions of families of functions in general position*, Uspekhi Mat. Nauk **37** (1982), no. 4, 129–130; English transl. in Russian Math. Surveys **37** (1982).
15. B. Morin, *Formes canoniques des singularités d'une application différentiable*, C. R. Acad. Sci. Paris **260** (1965), 5662–5665, 6503–6506.
16. A. I. Panasyuk, *On the dynamics of attainable sets of control system*, Differentsial'nye Uravneniya **24** (1988), no. 12, 2105–2115; English transl. in Differential Equations **24** (1988).

17. L. S. Pontryagin, V. G. Boltyanskiĭ, R. V. Gamkrelidze, and E. F. Mishchenko, *The mathematical theory of optimal processes*, "Nauka", Moscow, 1983; English transl. of the 1st edition, Wiley, New York, 1962; Macmillan, New York, 1964.
18. J. Écalle, *Théorie iterative: Introduction à la théorie des invariants holomorphes*, Ann. Inst. Fourier (Grenoble) **29** (1979), no. 1, 263–282.

Translated by G. PASECHNIK

Nice Dimensions and Their Generalizations in Singularity Theory

V. M. ZAKALYUKIN

Let Ω be the space of smooth proper maps $f : N^n \to P^p$ of smooth manifolds N^n and P^p. The RL-equivalence group, consisting of pairs (G_N, G_P) of diffeomorphisms in N^n and P^p, acts in Ω by the formula $f \mapsto G_P \circ f \circ G_N^{-1}$. A map f is said to be *stable* if all maps close to it in a suitable topology are equivalent to it.

In a series of papers [7], which stimulated a lot of further research, Mather found a set of pairs (n, p) of dimensions such that, for any N^n and P^p, the set of stable maps is dense in Ω. Such dimensions were said to be *nice*. In nice dimensions, any stable map f in the neighborhood of any point $x \in X$, i.e., any germ (f, x), can be reduced to one from a finite list of normal forms by means of local diffeomorphisms of N and P.

Which dimensions are nice and what is the structure of typical objects in "bad" dimensions? These are natural questions in many areas of singularity theory. It is our intention here to discuss these questions for maps.

§1. Properties of stable maps

1.1. The property of being stable or unstable depends only on local properties of the map.

Let $j^k(n, p)$ be the space of k-jets at zero of maps $f : (\mathbb{R}^n, 0) \to (\mathbb{R}^p, 0)$. It consists of segments of degrees $\leq k$ of the Taylor series of the components of f. The group of k-jets of origin-preserving diffeomorphisms of $\mathbb{R}^n$ and $\mathbb{R}^p$ acts algebraically on $j^k(n, p)$. Let $_sJ^k(N, P)$ be the space of s-multijets of f, that is, the jets defined at all s-tuples $x_1, \ldots, x_s$ of distinct points of N with arbitrary values in P. It breaks up into orbits of the RL-equivalence group, which are locally isomorphic to direct products of orbits in $j^k(n, p)$

and to s copies of N and P. Define $j^k f$, called the *jet extension* of f, to be the map that takes an s-tuple of points of N into the k-jet of f at those points.

THEOREM 1.1 [7, V]. *A proper map is stable if and only if for some (and hence for every) s and k such that $k \geq p$ and $s \geq p + 1$ the jet extension $j^k f$ is transversal to all orbits in $_s J^k(N, P)$.*

1.2. According to one version of the Transversality Theorem, a countable intersection of open dense subsets of Ω contains the set of maps f whose jet extensions are transversal to a given manifold in $_s J^k(N, P)$.

To simplify discussion, below we shall consider only local objects.

1.3. Consider the group of contact (V-)equivalences of map germs $f: (\mathbb{R}^n, 0) \to (\mathbb{R}^p, 0)$, whose elements are coordinate changes in the inverse image and multiplications by nonsingular $p \times p$ matrices with functions on $\mathbb{R}^n$ as elements. This group is often more convenient than the RL-equivalence group.

It turns out, that $j^k f$ is transversal to an RL-orbit if and only if it is transversal to the corresponding V-orbit, and RL-stability is equivalent to V-stability. The explanation for this at first glance surprising situation (since the group of V-equivalences is larger than the RL-group) is that the maps $j^k f$ are not typical maps from $\mathbb{R}^n$ to $j^k(n, p)$.

Let $C(x)$ be the ring of germs at zero of smooth functions of $x \in \mathbb{R}^n$, $M(x)$ the maximal ideal in $C(x)$, and $I = C(x)\{f_1, \ldots, f_p\}$ the ideal generated by the components of f. The local algebra $Q_f = C(x)/I$ is called the $\mathbb{R}$-algebra of f.

THEOREM [7, IV]. *Germs $(f_1, 0)$ and $(f_2, 0)$ are V-equivalent if and only if their $\mathbb{R}$-algebras are isomorphic.*

COROLLARY. *Define a genotype to be a map germ $\phi: x \mapsto y(x)$, $x \in \mathbb{R}^r$, $y \in \mathbb{R}^s$, with a zero 1-jet. By an unfolding of ϕ we mean a map of the form $f: (x, u) \mapsto (y(x), u)$, $u \in \mathbb{R}^k$. The germs ϕ and f have isomorphic $\mathbb{R}$-algebras. Any map is equivalent to an unfolding of some genotype; two maps are V-equivalent if and only if their genotypes are V-equivalent.*

§2. The region of nice dimensions

Suppose we are given an algebraic action of a group G on a finite-dimensional manifold U. Then U has a finite stratification [10] such that the action of G on each stratum S_i determines a smooth locally trivial bundle $\pi_i: S_i \to S_i/G$.

Define the set Π^m of m-modal orbits to be the closure of the union of all strata for which the base of the bundle π_i has dimension at least m. This set does not depend on the stratification.

Let $^m\Pi^k$ be the set of m-modal orbits of the group of k-jets of V-equivalences, which acts on $j^k(n,p)$, and let $^m\sigma^k(n,p)$ be the codimension of $^m\Pi^k$. It is clear that the inverse image of $^m\Pi^k$ under the projection $j^{k+1}(n,p) \to j^k(n,p)$ (which amounts to "forgetting" the coefficients of degree $k+1$ in the jet) is a subset of $^m\Pi^{k+1}$. Therefore, $^m\sigma^{k+1}(n,p) \le {}^m\sigma^k(n,p)$. Let $^m\sigma(n,p)$ denote the smallest $^m\sigma^k(n,p)$ for different k.

The complement of $^1\Pi^k$ consists of finitely many orbits. Consequently, if $n < {}^1\sigma(n,p)$, a small displacement of f will ensure that the map $j^k f$ of n-space into the space of jets will not intersect $^1\Pi^k$ and will be transversal to all orbits of the complement of $^1\Pi^k$. If $k \ge p$, the germ f is stable.

Let $n > {}^m\sigma(n,p)$ and assume that k is sufficiently large. The jet extension $j^k f$ of a typical germ f is transversal to the strata S_i. Consider one of the strata S, with modality $m = \dim(S/G)$ and codimension c. Consider the germ of the smooth manifold $S(f)$, i.e., the intersection of the image of $j^k f$ with S. The image $\pi(S(f))$ is an invariant of the germ under the V-equivalence group.

There are two possible cases:

(1) If $n - c < m$, the image $\pi(S(f))$ at a regular point of $\pi|_{S(f)}$ is the germ of the smooth $(n-c)$-dimensional manifold S/G, that is, it can be defined by a set of $m - n + c$ functions of $n - c$ variables in suitable local coordinates on S/G. It follows that in a neighborhood of f in Ω there exists a family of germs parametrized by a set of $m - n + c$ functions of $n - c$ variables that are not equivalent to each other for different functions. Such functions are called *functional moduli*.

(2) If $n - c \ge m$, then the map $j^k f$ is transversal to the V-orbit of $j^k f(x_0)$ at a regular point of $\pi|_{S(f)}$, corresponding to a k-jet $j^k f(x_0)$. Hence the map is transversal to all nearby orbits and the germ (f, x_0) is stable. For a typical f, however, the set of critical points of $\pi|_{S(f)}$ has dimension $n - c - m$ and the set of critical values of $\pi|_{S(f)}$ is an invariant of f.

Thus, the region of nice dimensions is determined by the condition $n < {}^1\sigma(n,p)$. Using the classification of $\mathbb{R}$-algebras, Mather [7, VI] constructed an appropriate stratification of $j^k(n,p)$ and computed $^1\sigma(n,p)$. The values of $^1\sigma$ are given in Table 1 on p. 140; for the region of nice dimensions, see Figure 1 on p. 140.

FIGURE 1

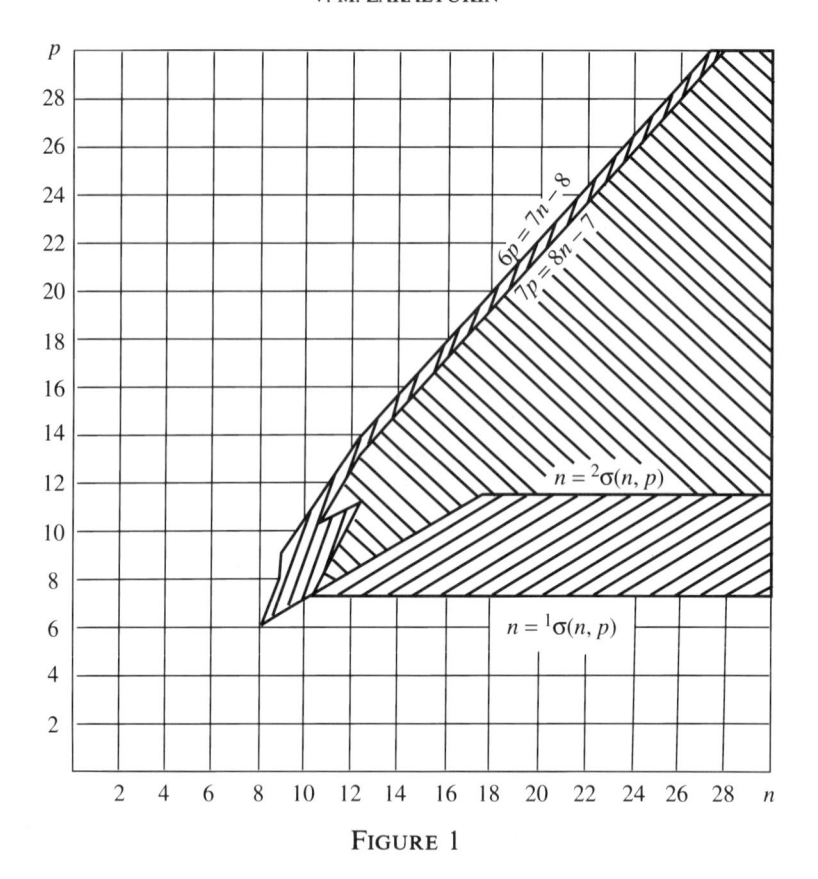

Values of	$\Delta = p - n$	$^1\sigma(n,p)$	$^2\sigma(n,p)$
	$\Delta \geq 4$	$8 + 6\Delta$	$7 + 7\Delta$
	$3 \geq \Delta \geq 1$	$9 + 6\Delta$	$16 + 4\Delta$
	0	9	13
	-1	9	11
	-2	8	13
	$-3 \geq \Delta \geq -7$	$7 - \Delta$	$-2\Delta + 4$
	$-7 \geq \Delta$	$7 - \Delta$	$11 - \Delta$

TABLE 1

§3. Seminice dimensions

A germ $(f, 0)$ is said to be *finitely determined* if there exists k such that the RL-orbit of $(f, 0)$ contains all germs with the same k-jet as f.

Theorem 1.1 implies that a stable germ is $(p+1)$-determined.

A set of germs, the codimension of whose k-jets in $j^k(n,p)$ increases [without bound] with increasing k, is called a set of infinite codimension.

THEOREM [9]. *For any n, p, the finitely determined germs for the contact group (singularities of finite type) form a set whose complement has infinite codimension.*

The situation for the RL-equivalence group is more complex.

THEOREM (du Plessis [4]). *For $n \leq {}^2\sigma(n,p)$, the set of not finitely determined germs has codimension ∞; for $n > {}^2\sigma(n,p)$, it has finite codimension.*

The values of ${}^2\sigma(n,p)$, computed by Wall [11], are listed in Table 1; the boundary $n = {}^2\sigma(n,p)$ of the set of seminice dimensions is shown in Figure 1.

The proof of du Plessis's theorem is based on the following result of Gaffney [5]: A germ f is finitely determined if and only if it has finite type and there exists a representative $\tilde{f}: U \to W$ of this germ, defined in some neighborhood U, such that for any $y \in W \setminus \{0\}$ the germ on the set $\tilde{f}^{-1}(y) \cap \Sigma_f$ is stable, where Σ_f is the set of critical points of f. The necessity of stability away from zero is due to existence of finitely many germs with zero k-jet at zero—which thus belong to the tangent space of the RL-orbit of $(f,0)$—certain representatives of which generate $C(x-x_0)/\mathfrak{M}^{p+1}(x-x_0)$ for all $x_0 \neq 0$ close to zero.

Consider the case $n \leq {}^2\sigma(n,p)$. The jet extension $j^k f$ of a typical germ intersects finitely many 0-modal orbits transversally, is transversal to finitely many strata of unimodal orbits, and intersects strata of greater modality at certain points $j^k f(x_i)$. The points of instability are points x_i and the critical points of maps $\pi_i|_{S_i(f)}$ for unimodal strata. Since these points are isolated, it follows from Gaffney's theorem that the germ is finitely determined.

When $n > {}^2\sigma(n,p)$, the appearance of functional moduli that depend on coefficients of arbitrarily high order in $j^k f$, or of sets of points at which the germs f are not stable, which have positive dimension in $\mathbb{R}^n$, precludes the existence of finite determinacy for almost all germs.

§4. Contact classes of 2-jets

The space $j^1(n,p) \approx \operatorname{Hom}(\mathbb{R}^n, \mathbb{R}^p)$ breaks up into classes Σ^r, $\max(0, n-p) \leq r \leq n$, where Σ^r is defined as the set of 1-jets f such that the kernel of df at zero has codimension r. The algebraic set Σ^r has codimension $r(p-n+r)$ in $j^1(n,p)$. All germs of the class Σ^0 are nonsingular and are equivalent to each other. Note that if $p \geq 2n$, then $\operatorname{codim} \Sigma^r > n$ for all $r > 0$; therefore, for such n and p typical map germs have no singularities.

Consider the differential $df_x : T_xN \to T_{f(x)}P$ at a point x and let K^r be its kernel and $C = T_xP/\operatorname{Im} df_x$ its cokernel. Then the second intrinsic derivative $d(df) : T_xN \to \operatorname{Hom}(K, C)$ is defined independently of the local coordinates; corresponding to $d(df)$ we have a bilinear map $d^2f : K \times K \to C$. The latter can be considered as a linear map $\tilde{d}^2f$ of the symmetric square $K \circ K$ into C. The group $\operatorname{GL}(K) \times \operatorname{GL}(C)$ acts in a natural way on the space of such maps, and its orbits are in one-to-one correspondence with the orbits of 2-jets of genotypes $\phi : K \to C$ under the action of the group V^2 of 2-jets of V-equivalences. By the unfolding theorem, the V-orbit of a germ f is uniquely determined by the V-orbit of the genotype and the rank $n-r$ of the first differential.

In addition to the Boardman classes $\Sigma^{r,d}$, which consist of the 2-jets for which the kernel of $d(df)|_K$ is d-dimensional, let us consider the finer partition into the classes $\Sigma^{r,(a)}$ of jets for which the kernel L of the linear mapping $\tilde{d}^2f$ is a-dimensional. The class $\Sigma^{r,(a)}$ is not empty only if

$$\max\left(0, \frac{r(r-1)}{2}+n-p\right) \leq a \leq \min\left(\frac{r(r+1)}{2}, p+r-n\right).$$

The codimension of $\Sigma^{r,(a)}$ in Σ^r can be computed by the formula for codimension of the set of linear operators from $\mathbb{R}^{r(r+1)/2}$ to $\mathbb{R}^{n-p+r}$ with a-dimensional kernel. Therefore, in the space $j^2(n,p)$ we have

$$\operatorname{codim} \Sigma^{r,(a)} = r(p-n+r) + a\left(a+p-n-\frac{r(r-1)}{2}\right).$$

The closure of $\Sigma^{r,(a)}$ consists of the classes $\Sigma^{r,(b)}$ with $b \geq a$; if $\Sigma^{r,(a)}$ is not empty, then $\operatorname{codim} \Sigma^{r,(b)} \geq \operatorname{codim} \Sigma^{r,(a)}$ for every $b \geq a$. Indeed, $\operatorname{codim} \Sigma^{r,(b)} - \operatorname{codim} \Sigma^{r,(a)} = (b-a)\left(a+b+p-n-\frac{r(r-1)}{2}\right)$ is positive in view of the above restrictions on a.

PROPOSITION. *The V-equivalence class of j^2f is determined by the orbit of L under the action of the group $\operatorname{PGL}(K)$ in the Grassmannian $\operatorname{Gr}\left(\frac{r(r+1)}{2}, a\right)$ of a-dimensional planes in $K \circ K$.*

The proof follows from the V-classification of 2-jets of maps according to their $\mathbb{R}$-algebras $Q_f^{(2)} = C(x)/I + M^3(x)$. Note that the ideal $I + M^3(x)$ is uniquely determined by L and $a = \dim_{\mathbb{R}} Q_f^{(2)}$.

Let $\alpha : \Sigma^{r,(a)} \to \operatorname{Gr}\left(\frac{r(r+1)}{2}, a\right)$ and $\alpha^c : \Sigma^{r,(a)} \to \operatorname{Gr}\left(\frac{r(r+1)}{2}, \frac{r(r+1)}{2}-a\right)$ be the maps that take j^2f into the kernel of $\tilde{d}^2f$ and, respectively, the image of the dual $(\tilde{d}^2f)^* : C^* \to K^* \circ K^*$ of $\tilde{d}^2f$ ($(\tilde{d}^2f)^*$ maps a linear form β on C into the quadratic form $\beta \circ d^2f$ on K). Both α and α^c induce one-to-one maps of the orbit spaces which preserve codimensions of sets of orbits. The homeomorphisms

$$\operatorname{Gr}\left(\frac{r(r+1)}{2}, a\right) / \operatorname{PGL}(r) \approx \Sigma^{r,(a)}/V^2$$

$$\approx \operatorname{Gr}\left(\frac{r(r+1)}{2}, \frac{r(r+1)}{2} - a\right) / \operatorname{PGL}(r)$$

show that the orbit spaces for a and $\frac{r(r+1)}{2} - a$ are duals.

We now determine the modality of some families of orbits.

1. If $a = 0$ or $a = \frac{r(r+1)}{2}$, there is just one contact class. The condition for $\Sigma^{r,(0)}$ to be nonempty is $p - n \geq \frac{1}{2}r(r-1)$.

2. If $a = 1$, we obtain a unique element of $K \circ K$ or, by duality, a quadratic form on K. Rank (and signature in the real-valued case) determines finitely many V-orbits.

3. If $a = 2$, we obtain the subspace spanned by two linearly independent quadratic forms f and g. Such pencils of forms are classified by their eigenvalues λ/μ, up to a projective transformation of the complex line. Here the complex numbers λ and μ are determined by the condition that the form $\lambda f + \mu g$ have rank less than r. In general position, one obtains r different points in CP^1 which have $r - 3$ cross ratios (provided $r > 3$), all invariants of orbits.

4. If $a \geq 3$ (by duality, we need only consider the case $a \leq \frac{1}{4}r(r+1) = $ {half the dimension of $K \circ K$}), the number of parameters necessary to describe the equivalence classes is at least

$$m_2(r, a) = \dim \operatorname{Gr}\left(\frac{r(r+1)}{2}, a\right) - \dim \operatorname{PGL}(r) = a\left(\frac{r(r+1)}{2} - a\right) - r^2 + 1.$$

If L is a subspace in general position, the transformations that preserve it form a subgroup $\operatorname{St}(L)$ of zero dimension, that is, $m_2(r, a)$ is the number of moduli of a general family of orbits in $\Sigma^{r,(a)}$.

Thus, the 0-modal cases are (r, a), $\left(r, \frac{r(r+1)}{2} - a\right)$, $a = 0, 1$; and if $r \leq 3$, also $(r, 3)$, $\left(r, \frac{r(r+1)}{2} - 3\right)$; the unimodal cases are $(4, 2)$, $(4, 8)$, $(3, 3)$.

§5. Computation of $^1\sigma(n, p)$ and $^2\sigma(n, p)$

Mather [7, VI] proved that all germs in the classes $\Sigma^{r,(0)}$ and $\Sigma^{r,(1)}$, except for a subset of infinite codimension, have no moduli.

A list of all simple (0-modal) germs may be found in [6]. Dimca and Gibson [2, 3] obtained a classification of unimodal germs Σ^2, $n = p$. It turns out that all germs of finite type in $\Sigma^{2,1}$ are unimodal. For $n < p$ this was proved in [11]. The case $\Sigma^{2,2}$ leads to a classification of linear families of cubic curves depending on one parameter. It can be shown that all unimodal germs lie in $\Sigma^{2,2,0}$.

Wall [12] considered the case $n > p$. Working from the lists of singularities of functions relative to groups of right changes [1], he obtained lists of unimodal V-classes of hypersurface singularities, i.e., mappings in Σ^{n-p+1}.

In the case $r = 4$, $p = 3$, the classification of 2-jets implies the existence of six moduli. Hence, unimodal germs of Σ^{n-p+k}, $k > 1$, appear only if $k = 2$. The cross-ratios of the eigenvalues of pencils of quadratic forms $j^2 f$ limit the search for unimodal classes to the cases $r = 3$ and $r = 4$. These classes were found using the Splitting Theorem [12], according to which a germ of a map can be expressed component-by-component as the sum of two germs, each depending on a smaller number of variables. The form of decomposition depends essentially on the Jordan form of the matrix that determines the eigenvalues of the pencil of quadratic forms.

The numbers $^1\sigma(n,p)$ and $^2\sigma(n,p)$ are the same in both complex- and real-valued cases; however, the existence of classes with different numbers of real and complex moduli (V. A. Vasil′ev) may lead, in general, to a difference between $^m\sigma_\mathbb{R}(n,p)$ and $^m\sigma_\mathbb{C}(n,p)$ in the holomorphic and real-analytic cases.

§6. Some estimates for the region $n \leq {}^m\sigma(n,p)$ for large m

PROPOSITION 6.1. *If a point (n,p) lies in the region $n > {}^m\sigma(n,p)$ of m-bad dimensions, then the ray $(n+l, p+l)$, $l \geq 0$, lies in the same region. The quantities ${}^m\sigma(n,p)$ on the ray $p - n = \text{const}$ stabilize from some n on.*

Note that when $n \geq 3$, $^1\sigma(n,p)$ and $^2\sigma(n,p)$ depend only on $n - p$.

PROOF. The stratification of $j(n+l, p+l)$ can be represented by the classes Σ^r, which for $r \leq n$ are unfoldings of Σ^r in $j^k(n,p)$, i.e., they have the same modality and codimension as in $j^k(n,p)$. The additional strata Σ^r, $r > n$, have codimensions that increase with increasing n.

PROPOSITION 6.2. ${}^m\sigma^2(n,n) = (2m)^{2/3}(1 + O(m^{-1/3}))$.

PROOF. If $\Delta = p - n = 0$, then the class Σ^r has codimension r^2 in $j^2(n,n)$. The minimal value a_0 for a nonempty class $\Sigma^{r,(a)}$ is $a_0 = \frac{r(r-1)}{2}$, and then the codimension of $\Sigma^{r,(a_0)}$ in Σ^r is zero. The modality of the 2-jets in general position belonging to $\Sigma^{r,(a_0)}$ is $m_2(r) = \frac{r^3}{2} - \frac{3r^2}{2} + 1$.

For a class $\Sigma^{r,(a)}$, $a = a_0 + b$, the codimension and modality of a jet in general position is determined by the formulas

$$c(r,b) = r^2 + b^2 + br\frac{r-1}{2},$$

$$m_2(r,b) = \left(\frac{1}{2}r(r-1) + b\right)(r-b) - r^2 + 1$$

$$= \frac{r^3}{2} - \frac{3r^2}{2} + 1 - b\left(\frac{r^2}{2} - \frac{3r}{2} + b\right).$$

If $r > 2$, then $\frac{r^2}{2} - \frac{3r}{2} + b > 0$. Therefore, as b increases, $c(r, b)$ increases and $m_2(r, b)$ decreases. Define $m = m_2(r)$. Then the minimal codimension $c(r)$ of classes for which $m_2(r, b) \geq m$ is r^2. Since $^m\sigma^2(n, n)$ is a nondecreasing function of m, its asymptotic behavior is determined by the sequence of values $m_2(r)$: $c(r) = r^2$, $m \sim r^3/2$.

In order to estimate $^m\sigma^2$, $m \to \infty$, we may restrict ourselves to the case of 2-jets in general position. Indeed, if z is a degenerate jet such that the dimension of the subgroup $\text{St}(z)$ is t, then the coefficients of the jet satisfy at least t independent conditions. Thus, the modality of a degenerate 2-jet in a class $\Sigma^{r,(a)}$ of codimension $c(r, a-a_0)+t$ is at most $m_2(r, a-a_0)+t$, that is, it increases with increasing codimension m slower than $m^{3/2}$.

We now consider the case of large $\Delta = p - n > 0$.

PROPOSITION 6.3. *The region* $n > {}^m\sigma(n, p)$ *of m-bad dimensions contains a ray* $p = \left(1 + \frac{1}{\alpha}\right)n + \beta$ *whose slope satisfies the estimate*

$$\alpha = \frac{3}{2}(4m)^{1/3}(1 + O(m^{-1/3})).$$

REMARKS. 1. Propositions 6.2 and 6.3 do not give sharp estimates for $^m\sigma(n, p)$ since when r, a are large, the k-jets, $k \geq 3$, contain many moduli that have not been considered here.

2. For $\Delta \geq 4$ the boundary of the region $n \leq {}^2\sigma(n, p)$ is the ray $p = \frac{8}{7}n - 1$, which emanates from the point $(35, 39)$, corresponding to the class $\Sigma^{4,(3)}$, which has six moduli. Consequently, this ray is also part of the boundary of the region $n \leq {}^m\sigma(n, p)$ for $m = 3, 4, 5, 6$.

PROOF OF PROPOSITION 6.3. The class $\Sigma^{r,(a)}$ is not empty provided that $a + \Delta \geq \frac{l^2-r}{2}$, in which case it has codimension $c(r, a) = (r + a)\Delta + r^2 + a^2 - \frac{ar(r-1)}{2}$ and the modality of the general orbit is $m_2(r, a) = \frac{r^2}{2}(a - 2) + \frac{1}{2}ar - a^2 + 1$. The set of pairs (n, p) of m-bad dimensions, $n > {}^m\sigma(n, p)$, contains the two-parameter family of straight lines $n = (r + a)\Delta + r^2 + a^2 - \frac{ar(r-1)}{2}$, where the parameters r and a satisfy the condition $m_2(r, a) \geq m$. It is not hard to see that the minimum slope $r + a$ for large m is attained when $r/a = 2 + O(m^{-1/3})$ and $r = (4m)^{1/3}(1 + O(m^{-1/3}))$.

In the other extreme case, when $n > p$ and $\Delta = n - p$ is large, the corank of df is $\Delta + r_1$, where $r_1 \geq 0$, and the class $\Sigma^{\Delta+r_1}$ has codimension $r_1(\Delta + r_1)$.

The class $\Sigma^{\Delta+1}$ of codimension $\Delta+1$ corresponds to $\mathbb{R}$-algebras generated by principal ideals. Therefore, it breaks into V-orbits of germs of functions $\phi : (\mathbb{R}^{\Delta+1}, 0) \to (\mathbb{R}, 0)$ with zero 1-jet. For functions $\phi \in \Sigma^{\Delta+1, l}$, that is, functions whose second differential has corank l, the generalized Morse Lemma [1] implies that there exists a coordinate change in $\mathbb{R}^{\Delta+1}$ that transforms ϕ to the form $\psi(x_1, \dots, x_l) \pm x_{l+1}^2 \pm \cdots \pm x_{\Delta+1}^2$, where $j^2\psi = 0$.

Thus, the orbits of 2-jets in $\Sigma^{\Delta+1}$ do not have moduli and the modality of a germ in $\Sigma^{\Delta+1,l}$ does not depend on Δ.

Let $\sigma(m)$ be the minimal codimension of an m-modal stratum in $\Sigma^{\Delta+1}$ in the space of k-jets of functions of $\Delta+1$ variables for sufficiently large k.

PROPOSITION 6.4. *For any m and any sufficiently large Δ, the boundary of the region $n =^m \sigma(n, p)$ of m-bad dimensions is a ray on the horizontal straight line $p = 1 + \sigma(m)$.*

PROOF. The codimension of the closure of $\Sigma^{\Delta+2}$ in the space $j^k(n, p)$, which contains all $\Sigma^{\Delta+r}$, $r \geq 2$, is at least $2(\Delta + 2)$; for sufficiently large Δ it exceeds $1 + \Delta + \sigma(m)$, the minimum codimension of an m-modal stratum in $\Sigma^{\Delta+1}$. Therefore, for sufficiently large Δ the equation of the boundary is $n = 1 + n - p + \sigma(m)$.

Arnol'd's fundamental classification of singularities of functions relative to the group of right changes [1], which involves all 0-, 1- and 2-modal strata, implies the following conclusions:

1. $\sigma(1) = 6$ and the critical class P_8 is contained in $\Sigma^{3,3}$; $\phi \in P_8$ is a cubic form in three variables in general position.

2. $\sigma(2) = \sigma(3) = \sigma(4) = 10$, the critical class O_{14} is contained in $\Sigma^{4,4}$;

$$\phi = x^3 + y^3 + z^3 + w^3 + (ax + by + cz + dw)^3 + xyzw, \quad a, b, c, d \in \mathbb{R},$$

is a cubic form in four variables in general position.

3. $\sigma(5) \leq \cdots \leq \sigma(9) \leq 16$; the class $V'' \subset \Sigma^{3,3,3}$ contains nondegenerate forms of degree four in three variables.

PROPOSITION 6.5. *Let $\sigma_{\mathbb{R}}(m)$ be the minimal codimension of an m-modal stratum in the space of jets of functions of sufficiently many variables relative to the group of right changes. Then*

$$\sigma_{\mathbb{R}}(m) = \frac{1}{2}(\log_2 m)^2 (1 + O(\log^{-1} m)).$$

PROOF. By a theorem of Varchenko [1], the modality of a general germ f of a class $\Sigma(\Gamma)$ corresponding to a Newton diagram Γ is equal to its internal modality, that is, to the number of monomials in a basis of the local ring $C(x)/C(x)\{\partial f/\partial x_i\}$ which lie on and above Γ. Note that $\operatorname{codim}\Sigma(\Gamma)$ is equal to the number of basis monomials below Γ.

For almost all functions with a fixed Newton diagram, one can compute the multiplicity $\mu(f) = \operatorname{codim} \Sigma + m(f)$ of a singular point by a theorem of Kouchnirenko:

$$\mu(f) = n!V - \sum_{i=1}^{n}(n-1)!V_i + \sum_{i,j}(n-2)!V_{ij} \cdots \pm 1,$$

where V is the volume of the n-dimensional domain in the positive orthant below the Newton diagram, V_i is the $(n-1)$-dimensional volume below the Newton diagram on the ith coordinate hyperplane, V_{ij} is the $(n-2)$-dimensional volume on the coordinate plane orthogonal to the ith and jth basis vectors, and so on. We are assuming that Γ has points on each of the coordinate axes. This can be achieved for any finitely determined function.

Consider the class $\Sigma^{\langle k \rangle l} = \Sigma^{k,k,\ldots,k}$ (l indices) of germs of functions of k variables with zero l-jet. The Newton diagram for a typical function of this class is a simplex. We have

$$\mu(k,l) = l^k, \quad \operatorname{codim} \Sigma^{\langle k \rangle l} = \binom{k+l}{l}, \quad m\left(\Sigma^{\langle k \rangle l}\right) = l^k - \binom{k+l}{l}.$$

The proof of the proposition follows from the following lemmas:

LEMMA 6.6. *For all Newton diagrams Γ of functions of k variables which satisfy the conditions of Kushnirenko's theorem and have multiplicity $\mu(\Gamma) = l^k$,*

$$\frac{\operatorname{codim}(\Gamma)}{\mu(\Gamma)} \geq \binom{k+l}{l} \cdot l^{-k}.$$

For diagrams on the plane the assertion of the lemma is evident. If we replace a diagram Γ containing points $(i, 0)$ and $(0, j)$ by the segment with ends at these points, the multiplicity increases by twice the area between the diagram and the segment, while the codimension increases by the area itself. Hence the minimum quotient $\operatorname{codim}(\Gamma)/\mu(\Gamma)$ can be obtained by considering only segments, that is, for diagrams of quasihomogeneous functions.

LEMMA 6.7. *Let $c(m) = \min\{\binom{k+l}{l} \mid l^k - \binom{k+l}{l} \geq m\}$. Then*

$$c(m) = \frac{1}{2}(\log_2 m)^2 (1 + O(\log^{-1} m)).$$

REMARKS. 1. The number of V-equivalence moduli in each class can be computed by using spectral sequences to reduce the functions to normal form [1] and to determine the dimensions of approximations to the homogeneous components of the ideal $C(x)\{\phi, \partial\phi/\partial x_i\}$.

2. For the Boardman class $\Sigma^{i_1,\ldots,i_n}$ in the space of map germs $f: \mathbb{R}^n \to \mathbb{R}^p$, let $m(i_1, \ldots, i_n)$ denote the modality of a general stratum. Most likely, for all (n, p), the asymptotic behavior of ${}^m\sigma(n, p)$, $m \to \infty$, is determined by $\min\{\operatorname{codim} \Sigma^{i_1,\ldots,i_n} \mid m(i_1, \ldots, i_n) > m\}$; the critical class is determined by the maximal values of i_1, i_2, and so on. In other words, the appearance of the maximal number of moduli for a given codimension corresponds to the maximal possible degeneration in lower-order jets.

References

1. V. I. Arnol'd, A. N. Varchenko, and S. M. Gusein-Zade, *Singularities of differentiable maps*, vol. I, "Nauka", Moscow, 1982; English transl., Birkhäuser, Basel, 1985.
2. A. Dimca and C. G. Gibson, *Contact germs from the plane to the plane*, Singularities, Part 1, Proc. Symp. Pure Math., vol. 40, Amer. Math. Soc., Providence, R. I., 1983, pp. 277–282.

3. _____, *Contact unimodal germs from the plane to the plane*, Quart. J. Math. **34** (1983), 281–295.
4. A. A. du Plessis, *Genericity and smooth finite determinacy*, Singularities, Part 1, Proc. Symp. Pure Math., vol. 40, Amer. Math. Soc., Providence, R. I., 1983, pp. 295–312.
5. T. Gaffney, *The structure of $T\mathscr{A}(f)$, classification and an application to differential geometry*, Singularities, Part 1, Proc. Symp. Pure Math., vol. 40, Amer. Math. Soc., Providence, R. I., 1983, pp. 409–427.
6. M. Giusti, *Classification des singularités isolées d'intersections complètes simples*, C. R. Acad. Sci. Paris Sér. I Math. **284** (1977), 167–170.
7. J. Mather, *Stability of C^∞ mappings*. I, Ann. of Math. **87** (1968), 89–104; II, Ann. of Math. **89** (1969), 254–291; III, Inst. Hautes Études Sci. Publ. Math. **35** (1969), 127–156; IV, Inst. Hautes Études Sci. Publ. Math. **37** (1970), 223–248; V, Adv. in Math. **4** (1970), 301–335; VI, Lecture Notes in Math., vol. 192, Springer-Verlag, Berlin, Heidelberg, and New York, 1971, pp. 207–253.
8. A. G. Kouchnirenko, *Polyèdres de Newton et nombres de Milnor*, Invent. Math. **32** (1976), 1–31.
9. J. C. Tougeron, *Idéaux des fonctions différentiables*, Ann. Inst. Fourier (Grenoble) **18** (1968), 177–240.
10. A. A. Rosenlicht, *A remark on quotient spaces*, An. Acad. Brasil Ciênc. **35** (1963), 487–489.
11. C. T. C. Wall, *Determination of the seminice dimensions*, Math. Proc. Cambridge Philos. Soc. **97** (1985), 79–88.
12. _____, *Classification of unimodal isolated singularities of complete intersections*, Singularities, Part 2, Proc. Symp. Pure Math., vol. 40, Amer. Math. Soc., Providence, R. I., 1983, pp. 625–640.

Translated by A. BOCHMAN

On an Inverse Problem of Measure Theory

V. M. KLIMKIN

The following theory, known in classical measure theory as Nikodým's Theorem, is a striking improvement of the uniform boundedness principle in the space $ca(T, \Sigma)$ (see [1, p. 309]). Let $\Phi \subset ca(T, \Sigma)$. If $\sup\{|\varphi(E)|, \varphi \in \Phi\} < \infty$ for every E in Σ, then
$$\sup\{|\varphi(E)|, \varphi \in \Phi, E \in \Sigma\} < \infty.$$

Bobynin [2] was apparently the first to prove Nikodým's Theorem using only measure theory. It was shown in [1] that Nikodým's Theorem is not true for a family of measures on an algebra of sets. It was shown in [10] that analogs of theorems in classical measure theory for set functions whose domain is not a ring are important in the study of certain problems in theoretical physics.

EXAMPLE 1.1. Let Σ be the class of all open subsets of the interval $(0, 1)$. Clearly, $0 \in \Sigma$, Σ is closed under the formation of countable unions, but Σ is not a ring. We denote $T = (0, 1)$, $E_n = \left(0, \frac{n}{n+1}\right)$, $n = 1, 2, \ldots$, and define set functions on Σ as follows:

$$\mu_1(E) = \begin{cases} 1, & \text{if } E = T, \\ 1, & \text{if } E \neq T \text{ and } E \supset E_1, \\ 0, & \text{otherwise}; \end{cases}$$

$$\mu_n(E) = \begin{cases} 1, & \text{if } E = T, \\ 1, & \text{if } E \neq T, E \supset E_{n-1}, \text{ and } E \not\supset E_n, \\ n, & \text{if } E \supset E_n, \\ 0, & \text{otherwise}. \end{cases}$$

It is easy to see that every μ_n, $n = 1, 2, \ldots$, is a countably additive set function, and for every set $E \in \Sigma$:
$$\sup\{\mu_n(E), n = 1, 2, \ldots\} < \infty.$$

On the other hand,
$$\sup\{\mu_n(E), E \in \Sigma, n = 1, 2, \ldots\} = \infty.$$

Therefore, Nikodým's Theorem is not true for Σ.

The following problem appears in classical measure theory in connection with Nikodým's Theorem.

Let Φ be a family of measures on an algebra of sets. We must have additional conditions, in terms of the family of measures or the algebra of sets, so that an analog of Nikodým's Theorem should be true. In this lecture we shall discuss the following problem, which is in a sense the inverse of the above problem.

Let $\Phi = \{\varphi\}$ be a family of scalar-valued set functions defined on a class of sets Σ. We want to establish conditions under which the functions in Φ will be uniformly bounded. This problem was first considered in [4], and, in a less general setting, by J. Dobrakov in [8]. Some applications of the problem, e.g., in classical measure theory, will be given later.

§1. Basic notation and definition

Let T be a set, Σ a class of subsets of T, $0 \notin \Sigma$. A sequence of pairwise disjoint sets $\{E_n\} \subset \Sigma$ will be called a *spectrum*. Any two disjoint sets $A, B \in \Sigma$ such that $A \cup B \in \Sigma$ will be called a *pair*. Σ will be called an *m*-class if it is closed under the formation of differences. Σ will be called a σ-class if $0 \in \Sigma$ and the union of any spectrum $\{E_n\} \subset \Sigma$ belongs to Σ.

Unless otherwise stated, we shall always assume that set functions φ under consideration are defined on Σ, take values in $\mathbb{R}^+$, and $\varphi(0) = 0$.

DEFINITION 1.1. We say that the functions in Φ have the *selection property* if, for any numbers $a > 0$ and $b > 0$, there exists $c > 0$ such that, for any $\varphi \in \Phi$ and any pair A, B in Σ, the inequalities $\varphi(A) < a$ and $\varphi(A \cup B) < b$ imply $\varphi(B) < c$. We say that functions in Φ have the *convexity property* if, under the same conditions, $\varphi(A) < a$ and $\varphi(B) < b$ imply $\varphi(A \cup B) < C$.

EXAMPLE 1.2. Let $\{\varphi_\alpha\}_{\alpha \in J}$ be a family of additive set functions defined on Σ with values in a normed space X.

For all $\alpha \in J$ and $E \subset \Sigma$, define
$$\mu_\alpha(E) = \|\varphi_\alpha(E)\|,$$
$$\nu_\alpha(E) = \exp(\|\varphi_\alpha(E)\|) - 1,$$
$$\eta_\alpha(E) = \sum_{k=1}^{n} c_k \|\varphi_\alpha(E)\|^k, \qquad c_k \geq 0.$$

Set functions in the families $\{\mu_\alpha\}_{\alpha \in J}$, $\{\nu_\alpha\}_{\alpha \in J}$, and $\{\eta_\alpha\}_{\alpha \in J}$ have both selection and convexity properties.

DEFINITION 1.2. Let S be a class of sequences in Σ. We shall say that functions in Φ are

(1) *bounded on sequences from S* if, for every $\{E_n\} \in S$,
$$\sup\{\varphi(E_n), \varphi \in \Phi, n = 1, 2, \ldots\} < \infty;$$

(2) *weakly uniformly bounded on sequences from S* if there exists $M > 0$

such that for every $\{E_n\} \in S$ and every function $\varphi \in \Phi$, there is a number n_0 (which depends on the function and the sequence) such that $\varphi(E_n) < M$ for all $n > n_0$;

(3) *weakly bounded on sequences from* S if for any $\{E_n\} \in S$ there exists $M > 0$ such that, for every $\varphi \in \Phi$, there is a number n_0 (which depends on the function) such that $\varphi(E_n) < M$ for all $n > n_0$.

Later we will consider the special cases in which S is the family of spectra in Σ, the family of increasing sequences in Σ, etc.

DEFINITION 1.3. We shall say that functions in Φ are

(1) *pointwise bounded* if, for any set E,

$$\sup\{\varphi(E), \varphi \in \Phi\} < \infty;$$

(2) *uniformly bounded*, if

$$\sup\{\varphi(E), \varphi \in \Phi, E \in \Sigma\} < \infty.$$

DEFINITION 1.4. A class of sets Σ is said to have the f_1-*property* (cf. [6]), if, for any spectra $\{E_n\}$ and $\{F_n\}$ in Σ satisfying the condition $E_n F_k \neq \varnothing$ for all $n, k \in N$, there exist an infinite set of numbers $\mathscr{P} \subset N$ and a set $F \in \Sigma$ such that $F_k \subset F$ for all $k \in \mathscr{P}$ and $F_k F = E_n F = \varnothing$ for all $n \in N$ and $k \notin \mathscr{P}$.

EXAMPLE 1.3. We will construct an algebra of sets with the f_1-property which is not a σ-algebra.

Let $\widehat{A} = N/\Delta$, where Δ is the ideal of all finite sets of natural numbers. Let $\{\widehat{E}_n\}$ and $\{\widehat{F}_n\}$ be spectra in A such that $\widehat{E}_n \widehat{F}_k = \varnothing$ for all $n, k \in N$. For each $n \in N$ fix $E_n \in \widehat{E}_n$ and $F_n \in \widehat{F}_n$ and define $D_1 = F_1$, $C_1 = E_1 \setminus F_1$,

$$D_n = F_n \setminus \left(\left(\bigcup_{k=1}^{n-1} F_k\right) \cup \left(\bigcup_{k=1}^{n-1} E_k\right)\right), \quad n = 2, 3, \ldots,$$

$$C_n = E_n \setminus \left(\left(\bigcup_{k=1}^{n-1} F_k\right) \cup \left(\bigcup_{k=1}^{n-1} E_k\right)\right), \quad n = 2, 3, \ldots.$$

We obtain spectra $\{D_n\}$ and $\{C_n\}$ in 2^N, which satisfy the condition $D_k C_n = \varnothing$ for all $k, n \in N$. Obviously, $C_n \in \widehat{E}_n$ ($D_n \in \widehat{F}_n$) for $n \in N$, since C_n and E_n (D_n and F_n) differ by a finite set. Define

$$F = \bigcup_{n=1}^{\infty} D_n$$

and consider the corresponding class $\widehat{F} \in \widehat{A}$. It is clear that $\widehat{F} \geq \widehat{F}_n$ and $\widehat{F} \wedge \widehat{E}_n = \varnothing$, $n = 1, 2, \ldots$. Consequently, the Stone space of the Boolean algebra $\widehat{A}$ is an algebra of sets Σ with the f_1-property.

Now we show that Σ is not a σ-algebra. Let $\{E_n\}$ be a spectrum of infinite sets in Σ. We are going to show that the spectrum $\{\widehat{E}_n\}$ in $\widehat{A}$ does not have the least upper bound in $\hat{A}$. Let $\hat{E} \in \hat{A}$ and suppose that $\hat{E} \geq \hat{E}_n$ for all $n = 1, 2, \ldots$. Choose $E \in \hat{E}$ and, for every $n \in N$, fix a natural number $m_n \in E_n E$. We now have a sequence $C = \{m_n\}$ of distinct natural numbers. Define $A = F \setminus C$. Clearly, A is an upper bound for $\{\hat{E}_n\}$ which is strictly less then $\hat{E}$. Consequently, $\bigcup_{n=1}^{\infty} E_n \notin \Sigma$.

Notice that there exist rings of sets with the f_1-property that are not σ-rings.

We now state some properties of classes of sets with the f_1-property.

PROPOSITION 1.1. *If Σ has the f_1-property, then for any spectrum $\{E_n\} \subset \Sigma$ there exist a sequence of infinite sets $\{N_k\}$ of natural numbers and a spectrum $\{F_i\} \subset \Sigma$ such that $N_k N_m \neq \varnothing$ for all $m \neq k$ and $E_n \subset F_k$ for all $n \in N_k$, $k = 1, 2, \ldots$.*

PROPOSITION 1.2. *Let Σ be an m-class that has the f_1-property, $\{E_n\}$ a spectrum in Σ, and $\{N_k\}$ a decreasing sequence of infinite subsets of natural numbers such that*

$$n_k = \min \mathscr{P} \notin \mathscr{P}_{k+1}, \qquad k = 1, 2, \ldots.$$

Then for any decreasing sequence of sets $\{F_k\} \subset \Sigma$ such that $E_n \subset F_k$ for all $n \in N_k$, $k = 1, 2, \ldots$ there exist a sequence of numbers $\{k_j\}$ and a set $F \in \Sigma$ such that $F_{n_{k_j}} \subset F$ and

$$F \setminus \left(\bigcup_{p=1}^{j} E_{n_{k_p}} \right) \subset F_{k_{j+1}}, \qquad j = 1, 2, \ldots.$$

Main results.

THEOREM 1.1. *The set functions in a family Φ, defined on an m-class Σ, are uniformly bounded if and only if one of the following conditions is satisfied:*

(1) *functions in Φ are bounded on the spectra in Σ and have the selection and convexity properties;*

(2) *functions in Φ are bounded on the spectra in Σ, bounded on the decreasing sequences in Σ, and have the convexity property.*

REMARK 1.1. Each of the requirements in either condition is essential for the theorem to hold.

COROLLARY 1.1.1. *Functions in a family Φ of functions defined on a ring Σ are uniformly bounded if and only if they are bounded on increasing sequences in Σ and have the selection property.*

COROLLARY 1.1.2. *Functions in a family Φ of functions defined on an algebra Σ or a σ-ring S are uniformly bounded if and only if the following conditions are satisfied:*

(1) *functions in Φ have the selection property;*

(2) *set functions in* Φ *are either bounded on spectra in* Σ *or bounded on monotone sequences in* Σ.

The following theorems specify conditions under which functions in Φ are bounded on spectra in Σ.

THEOREM 1.3. *If functions in a family* Φ *of functions defined on a ring* Σ *are weakly uniformly bounded on spectra in* Σ, *weakly bounded on increasing sequences in* Σ, *and have the selection property, then they are bounded on spectra in* Σ.

REMARK 1.2. If the class Σ in Theorem 1.3 is an algebra, the theorem remains true if the words "increasing sequences" are replaced by "decreasing sequences".

THEOREM 1.4. *Let* Σ *be either a* σ-*class or an m-class with the* f_1-*property. If functions in a family* Φ *of functions defined on* Σ *are weakly uniformly bounded on spectra, pointwise bounded and have the selection property, then they are uniformly bounded on spectra in* Σ.

Corollary 1.1.2 and Theorem 1.4 imply the following proposition which is in a sense an inverse of Nikodým's Theorem.

COROLLARY 1.4.1. *Set functions in a family* Φ *of functions defined on a* σ-*ring are uniformly bounded if and only if the following conditions are satisfied*:
  (1) *elements from* Φ *have the selection property*;
  (2) *elements from* Φ *are uniformly bounded on spectra in* Σ;
  (3) *for every* $E \in \Sigma$, $\sup\{\varphi(E), \varphi \in \Phi\} < \infty$.

We now apply the results of this section to vector-valued set functions (set functions with values in a topological abelian group).

Let (X, η) be a topological abelian group (TAG) where η is a base of symmetric neighborhoods of the zero element.

A set A in X is said to be *bounded in Bourbaki's sense* [3] if for any neighborhood $\nu \in \eta$ there exist a finite set $\{x_1, \ldots, x_n\} \subset X$ and a number $n \in N_n$ such that

$$A \subset \bigcup_{k=1}^{n} \{x_k + n\nu\},$$

where

$$n\nu = \left\{\sum_{k=1}^{n} y_k, \; y_k \in \nu\right\}.$$

A set A in X is said to be *bounded in Darst's sense* [7] if for any neighborhood $\nu \in \eta$ there exists a number $n \in N$ such that $A \subset n\nu$.

THEOREM 1.5. *Let functions in $\Phi = \{\varphi\}$ be defined on an m-class Σ and take values in a TAG (X, η). Then the set $\{\varphi(E), \varphi \in \Phi, E \in \Sigma\}$ is bounded in the sense of Darst (Bourbaki's sense) if and only if the following conditions are satisfied:*

(1) *for any continuous (finite and continuous) prenorm ρ on (X, η), the functions of the family $\{\rho \circ \varphi, \varphi \in \Phi\}$ ([1]) have the selection and convexity properties ([2]);*

(2) *for any spectrum $\{E_n\} \subset \Sigma$, the set*

$$\{\varphi(E_n), \varphi \in \Phi, n = 1, 2, \ldots\}$$

is bounded in the appropriate sense.

Now let (X, η, k) be a topological vector group [5] in a TAG. A set $A \subset X$ is said to be *C-bounded* if, for any neighborhood $\nu \in \eta$, there exist a finite set $M \subset X$ and a number $n \in N$ such that $A \subset \operatorname{co} M + n\nu$, where $\operatorname{co} M$ is the convex hull of M.

THEOREM 1.6. *Let the functions in Φ be defined on an m-class Σ with values in a locally convex topological vector group (X, η, k). The set*

$$\{\varphi(E), \varphi \in \Phi, E \in \Sigma\}$$

is C-bounded if and only if the following conditions are satisfied:

(1) *for any linear functional $f \in X'$, the functions in the family $\{f \circ \varphi, \varphi \in \Phi\}$ have the selection property and convexity properties;*

(2) *for any spectrum $\{E_n\} \subset \Sigma$, the set*

$$\{\varphi(E_n), \varphi \in \Phi, n = 1, 2, \ldots\}$$

is C-bounded.

REMARK 1.3. Other results in this section may be formulated in the same way.

§2. Uniform boundedness of regular set functions

There is a special branch of classical measure theory that deals with measures defined on Borel σ-algebras of topological spaces, in which context one imposes the requirement of pointwise boundedness, not on the whole σ-algebra, but only on the part of it consisting of the open sets. The following theorem is due to Dieudonné [9].

Let T be a compact Hausdorff topological space and Φ a family of regular Borel measures on the σ-algebra Σ of all Borel subsets of T. Assume that for any open set U we have

$$\sup\{\varphi(U), \varphi \in \Phi\} < \infty.$$

([1]) $\rho \circ \varphi(E) = \rho(\varphi(E))$, $E \in \Sigma$.

([2]) This condition is automatically satisfied for additive set functions (see, e.g., [2]).

Then
$$\sup\{|\varphi|(T), \varphi \in \Phi\} < \infty,$$
where $|\varphi|$ is the total variation of the measure φ.

In this section the problem stated in the previous section will be solved for regular set functions.

Let τ be the class of subsets of a set T, which is assumed to be closed under the formation of countable unions and finite intersections; let $\varnothing \in \tau$, $T \in \tau$. Following A. D. Alexandrov, we call (T, τ) a σ-*topological space*; sets in τ are called open sets and their complements closed sets. Let S and φ be classes of closed sets in the σ-topological space (T, τ); then S and φ are said to be separable if, for any disjoint sets $A \in S$ and $B \in \varphi$ there exist disjoint open sets U_1 and U_2 such that $A \in U_1$ and $B \in U_2$. Unless otherwise stated, it will always be assumed that set functions we consider are defined on an algebra Σ of subsets of the set T with $\Sigma \subset \tau$ and that they take values in $\overline{\mathbb{R}}^+$. Given a set function φ we define

$$\widetilde{\varphi}(E) = \sup\{\varphi(A), A \in E, A \in \Sigma\}, \qquad E \subset T.$$

DEFINITION 2.1. Let $M \subset \Sigma$, $\varepsilon > 0$; let $S \subset \Sigma$ be a class of closed sets. A function φ is said to be εS-regular on M if, for any $E \subset M$, there exists $F \in S$, $F \subset E$, such that $\widetilde{\varphi}(E\setminus F) \subset \varepsilon$. If only $\varphi(E\setminus F) \subset \varepsilon$, φ is said to be weakly εS-regular on M.

THEOREM 2.1. *Let (T, τ) be a σ-topological space and Σ an algebra of subsets of T, $\Sigma \supset \tau$. Let S and φ be τ-separable classes of closed sets in (T, τ). Let $\Phi = \{\varphi\}$ be a family of set functions defined on Σ with values on $[0, +\infty]$ all being weakly εS-regular on Σ_n and weakly $\varepsilon\varphi$-regular on τ.*

Then functions from $\Phi = \{\varphi\}$ are uniformly bounded on Σ if and only if the following conditions are satisfied:

(1) *for any $U \in \tau$ there exists a number $L(U)$ such that for any function $\varphi \in \Phi$, there is a finite collection of pairwise disjoint sets $\{D_1, \ldots, D_{k_\varphi}\}$ in $S \cup \varphi$ such that $D_p \subset U$, $p = 1, \ldots, k_\varphi$, and*

$$\widetilde{\varphi}\left(U \setminus \bigcup_{p=1}^{k_\varphi} D_p\right) < L(U);$$

(2) *functions in Φ have the selection property*;
(3) *functions in Φ are bounded on spectra of open sets*.

REMARK 2.1. If functions in Φ are ε-regular on τ with respect to either S or φ, condition (1) in Theorem 2.1 may be omitted.

REMARK 2.2. If S is the class of all closed sets, condition (1) in Theorem 2.1 may be omitted.

REMARK 2.3. By Theorem 1.4, condition (3) in Theorem 2.1 is equivalent to the following conditions:

(3′) *functions in Φ are uniformly bounded on spectra of open sets*;

($3''$) functions in Φ are pointwise bounded on open sets.

From Theorem 2.1 one easily derives the following theorem, which is the inverse in our sense of Dieudonné's theorem.

THEOREM 2.2. *Let (T, τ) be a Hausdorff topological space, Σ an algebra of subsets of T, $\Sigma \supset \tau$, and $S \subset \Sigma$ a class of compact subsets. Let functions in Φ, which are defined on Σ with values in $\overline{R}^+$, be εS-regular on Σ.*

Then the functions in Φ are uniformly bounded on Σ if and only if the following conditions are satisfied:

(1) *functions in Φ have the selection property*;
(2) *functions in Φ are pointwise bounded on open sets.*

We now apply results of this section to group-valued set functions.

A set function φ defined on an algebra Σ with values in a TAG (X, η) is said to be *regular on $M \subset \Sigma$ with respect to a class S* if, for any neighborhood $\nu \in \eta$ and any set $E \subset M$, there exists a set $F \in S$, $F \subset E$ such that $\varphi^{\vee}(E \setminus F) \subset \nu$. (If only $\varphi(E \setminus F) \subset \nu$, the word "regular" in this definition should be replaced by "weakly regular".)

THEOREM 2.3. *Let (T, τ) be a topological space, Σ an algebra of subsets of T, $\Sigma \supset \tau$, and let S and φ be τ-separable classes of closed sets. Let $\Phi = \{\varphi\}$ be a family of set functions defined on Σ with values in a TAG (X, η), which is weakly regular on Σ with respect to S and regular on τ with respect to φ.*

Then the set

$$\{\varphi(E), \varphi \in \Phi, E \subset \Sigma\} \qquad (1)$$

is bounded in Darst's (Bourbaki's) sense if and only if the following conditions are satisfied:

(1) *for any continuous (continuous and finite) prenorm ρ on (X, η), the functions in the family $\{\rho \circ \varphi, \varphi \in \Phi\}$ have the selection property*;
(2) *for any spectrum $\{U_n\} \subset \tau$ the set $\{\varphi(U_n), \varphi \in \Phi, n = 1, \ldots\}$ is bounded in the appropriate sense.*

REMARK 2.4. If, under the conditions of Theorem 2.3, the set functions take values in a locally convex topological vector group (X, η, k), there is an analogous criterion for C-boundedness of the set (1), with condition (1) replaced by the following condition:

($1'$) for any functional $f' \in X$, the functions in $\{f \circ \varphi, \varphi \in \Phi\}$ have the selection property.

THEOREM 2.4. *Let (T, τ) be a Hausdorff space, Σ an algebra of subsets of T, $\Sigma \supset \tau$, and $S \subset \Sigma$ a class of compact subsets. Let Φ be a family of set functions defined on Σ with values in a TAG (X, η), which are regular on Σ with respect to S.*

Then the set (1) *is bounded in Darst's* (*Bourbaki's*) *sense if and only if the following conditions are satisfied*:

(1) *for any continuous* (*continuous and finite*) *prenorm* ρ *on* (X, η), *the functions in the family* $\{\rho \circ \varphi, \; \varphi \in \Phi\}$ *have the selection property*;
(2) *for every set* $U \in \tau$ *the set* $\{\varphi(u), \; \varphi \in \Phi\}$ *is bounded in the appropriate sense.*

References

1. N. Dunford and J. T. Schwartz, *Linear operators*, vol. I, Interscience, New York, 1958.
2. M. N. Bobynin, *On a theorem in the theory of completely additive set functions*, Uspekhi Math. Nauk **7** (1952), no. 3, 113–120. (Russian)
3. N. Bourbaki, *Éléments de mathématique*, III. *Topologie générale*, 3-éme ed., Hermann, Paris, 1953.
4. V. M. Klimkin, *The uniform boundedness principle in generalized measure theory*, Dokl. Akad. Nauk SSSR **295** (1987), no. 4, 790–792; English transl. in Soviet Math. Dokl **36** (1988).
5. D. A. Raĭkov, *On B-complete topological vector groups*, Studia Math. **31** (1968), 295–306. (Russian)
6. D. Candelorn, *Alcuni teoremi di uniforme limitatazza*, Rend. Accad. Naz. Sci. XL Mem. Mat. **9(11)** (1985), 249–260.
7. R. B. Darst, *On a theorem of Nikodým with applications to weak convergence and Neumann algebras*, Pacific J. Math. **23** (1967), 473–477.
8. I. Dobrakov, *Uniform boundedness principle for exhaustive set functions*, Comment. Math. Prace Mat. **24** (1984), 202–205.
9. J. Dieudonné, *Sur la convergence des suites de mesures de Radon*, Ann. Accad. Brasil. Sci. **23** (1951), 21–38, 277–282.
10. S. Gudder, *Generalized measure theory*, Found. Phys. **3** (1978), 359–411.

Translated by V. OPERSTEIN

Classes of Coefficients of Convergent Random Series in Spaces $L_{p,q}$

S. YA. NOVIKOV

§1. Introduction. Preliminaries

The spaces $L_{p,q}$, familiar from interpolation theory for linear operators, have recently been attracting the attention of probability theoreticians. These spaces enable one to establish sharp convergence rates in classical limit theorems (see, e.g., [4]); in addition, they crop up naturally in connection with some traditional probability theoretic objects. For example, $\ell_{p,\infty}$ spaces appear in [5], which is devoted to the study of random series in L_p. The duality to be discussed in this paper was apparently first noticed in [5].

We recall the basic definitions. Let (W, Σ, μ) be a space with σ-finite measure, $L_0(\mu)$ the space of measurable μ-almost everywhere (a.e.) finite real-valued functions, $\eta(f; t) = \mu(|f| > t)$, $f^*(t) = \inf\{\alpha \geq 0 : \eta(f; \alpha) \leq t\}$, $t \geq 0$ the monotone rearrangement of a function f. The space $L_{p,q}$, $0 < p < \infty$, $0 < q < \infty$, consists of all functions $f \in L_0(\mu)$ such that

$$\|f\|_{p,q} = \left\{ \frac{q}{p} \int_0^\infty [t^{1/p} f^*(t)]^q \frac{dt}{t} \right\}^{1/q} < \infty.$$

The space $L_{p,\infty}$ is the set of all $f \in L_0(\mu)$ such that

$$\|f\|_{p,\infty} = \sup_{t>0} t^{1/p} f^*(t) < \infty.$$

If $p = q$, one gets the classical Lebesgue spaces L_p. In the general case the $L_{p,q}$ spaces are quasinormed and complete; for $p > 1$ and $q \geq 1$ they are Banach spaces.

For $W = \mathbb{N}$ and $W = \mathbb{Z}$, with the natural discrete measures, one obtains spaces of sequences of real numbers, denoted by $\ell_{p,q}$ and $\ell_{p,\infty}$. If $(W, \Sigma, \mu) = (\Omega, F, \mathbb{P})$ is a probability space, the $L_{p,\infty}$ spaces may be considered as the spaces of random variables (r.v.'s); in this case the measure $\mathbb{P}$ is assumed to be nonatomic, i.e., $(\Omega, F, \mathbb{R})$ is isomorphic as a measure

©1992 American Mathematical Society
0065-9290/92 $1.00 + $.25 per page

space to the unit interval with Lebesgue measure. There always exist continuous imbeddings $L_{p,q_1} \subset L_{p,q_2}$, $0 < p < \infty$, $0 < q_1 \leq q_2 \leq \infty$. If the measure space is a probability space, then $L_{s,\infty} \subset L_{p,q} \subset L_{r,1}$ for $r < p < s$, $0 < q \leq \infty$. For all these facts, the spaces $L_{p,q}$ are well known (see, e.g., [1, 3]). For $f \in L_0(W, \mu)$ denote $\rho_j(f) = \mu(2^j < |f| \leq 2^{j+1})$, $\sigma_j(f) = \mu(|f| > 2^j)$, $j \in \mathbb{Z}$.

LEMMA 1.1 (A. M. Shteinberg). *There exist positive constants $m = m(p, q)$ and $M = M(p, q)$ such that for any function $f \in L_0(W, \mu)$ we have*

$$(m\|f\|_{p,q})^p \leq \|\{2^{jp}\rho_j(f)\}\|_{q,p} \leq \|\{2^{jp}\sigma_j(f)\}\|_{q/p} \leq (M\|f\|_{p,q})^p,$$

i.e.,

$$\|f\|_{p,q} \asymp \|\{2^{jp}\rho_j(f)\}\|_{q/p}^{1/p} = \left[\sum_{j=-\infty}^{\infty} 2^{jq}\rho_j^{q/p}(f)\right]^{1/q}.$$

Letting $(W, \Sigma, \mu) = (\Omega, F, \mathbb{P})$, we now adopt the terminology of probability theory. Let $\Phi(p, q)$ denote the set of sequences $\{\xi_j\}$ of independent symmetric identically distributed r.v.'s, $\xi_1 \in L_{p,q}$. We consider the following classes of real sequences:

$$A^\infty(p, q) = \{a = (\alpha_k) \in \mathbb{R}^\infty : \sup|\alpha_k\xi_k| < \infty \quad \text{almost surely (a.s.)}$$
$$\text{for all } \{\xi_k\} \in \Phi(p, q)\};$$

$$A^0(p, q) = \{a = (\alpha_k) \in \mathbb{R}^\infty : \sum \alpha_k\xi_k \text{ converges a.s.}$$
$$\text{for all } \{\xi_k\} \in \Phi(p, q)\};$$

$$A(p, q) = \{a = (\alpha_k) \in \mathbb{R}^\infty : \sum \alpha_k\xi_k \text{ converges in } L_{p,q}$$
$$\text{for all } \{\xi_k\} \in \Phi(p, q)\}.$$

These classes turn out to be useful for a probabilistic characterization of p-summing operators. A description of such classes was obtained in [5]. It follows immediately from the definitions that $A(p, q) \subset A^0(p, q) \subset A^\infty(p, q)$.

§2. The classes $A^\infty(p, q)$

The following theorem of Banach about sublinear operators turns out to be useful in the study of these classes.

THEOREM 2.1 (Banach, 1926). *Let X be a complete metrizable topological vector space, $T_k : X \to L_0(\Omega, F, \mathbb{P})$, $k = 1, 2, \ldots$, a sequence of continuous linear operators. If $\sup_k |T_k x| < \infty$ a.s. for any $x \in X$, then the sublinear operator $x \to \sup_k |T_k x|$ acts continuously from X to $L_0(\Omega, F, \mathbb{P})$.*

We use this theorem to prove the following lemma.

LEMMA 2.2. *If $\sup_i |\alpha_i\xi_i| < \infty$ a.s. for all $\{\xi_i\} \in \Phi(p, q)$, then there exists $\varepsilon_0 > 0$ such that, for all $\xi \in L_{p,q}$ with $\|\xi\|_{p,q} \leq \varepsilon_0$, the series $\sum_{i=1}^\infty \mathbb{P}(|\alpha_i\xi| > 1)$ converges.*

PROOF. Define a new probability space $(\Omega_1, F_1, \mathbb{P}_1) = (\Omega^\infty, F^\infty, \mathbb{P}^\infty)$ and consider the sequence of linear operators $T_k \xi(\omega_1, \omega_2, \ldots) = \xi(\omega_k)$, acting from $L_{p,q}(\Omega)$ to $L_0(\Omega_1)$. Note that $\{T_k \xi\}$ is a sequence of identically distributed r.v.'s on Ω_1. Since the probability spaces $(\Omega, F, \mathbb{P})$ and $(\Omega_1, F_1, \mathbb{P}_1)$ are isomorphic, it follows that $\sup_i |\alpha_i T_i \xi| < \infty$ a.s. for all $\xi \in L_{p,q}$. By Banach's theorem, the sublinear operator $\xi \to \sup |\alpha_i T_i \xi|$ maps $L_{p,q}(\Omega)$ continuously into $L_0(\Omega_1)$. Consequently, there exists $\varepsilon_0 > 0$ such that for all ξ, $\|\xi\| \leq \varepsilon_0$,

$$\mathbb{P}_1(\sup_i |\alpha_i T_i \xi| > 1) < 1.$$

Elementary arguments now imply that the series $\sum_{k=1}^\infty \mathbb{P}(|\alpha_k \xi| > 1)$ converges, proving the lemma.

We can now prove the main result of this section.

THEOREM 2.3. *Let $a = \{\alpha_i\}$ be a real sequence, $0 < p < \infty$, $0 < q \leq \infty$. The following statements are equivalent*:

1) $a \in A^\infty(p, q)$;
2) *for any sequence $\{\xi_i\}$ of identically distributed r.v.'s in $L_{p,q}(\Omega)$,*

$$\lim_i \alpha_i \xi_i = 0 \quad a.s.;$$

3) *for any sequence $\{\xi_i\}$ of identically distributed r.v.'s in $L_{p,q}(\Omega)$,*

$$\sup |\alpha_i \xi_i| \in L_{p,\infty};$$

4) $a \in \ell_{p,r}$, *where r is defined by the relation $r^{-1} + q^{-1} = p^{-1}$ if $p < q$, and $r = \infty$ if $p \geq q$.*

PROOF. $2) \Rightarrow 1)$, $3) \Rightarrow 1)$ are clear. To prove $4) \Rightarrow 2)$ and $4) \Rightarrow 3)$, take an arbitrary $\lambda > 0$ and denote $r = [\log_2 \lambda]$, where $[\cdots]$ denotes the integral part of a number. We have

$$\sum_{i=1}^\infty \mathbb{P}(|\alpha_i \xi| > \lambda) \leq \sum_{i=1}^\infty \mathbb{P}(|\alpha_i \xi_i| > 2^r) \leq \sum_{j=-\infty}^\infty \rho_j(a) \mathbb{P}(2^{j+1}|\xi_1| > 2^r)$$

$$= \sum_{j=-\infty}^\infty \rho_j(a) \sigma_{r-j-1}(\xi_1) \leq \left(\frac{4}{\lambda}\right) \|\{2^{jp} \rho_j(a)\}\|_{r/p} \|\{2^{jp} \sigma_j(\xi_1)\}\|_{q/p}$$

(Hölder inequality)

$$\leq \left(\frac{4M \|a\|_{p,r} \|\xi_1\|_{p,q}}{\lambda}\right)^p \quad \text{(Lemma 1.1)}.$$

This estimate implies that, first, $\sup |\alpha_i \xi_i| \in L_{p,\infty}$ since

$$\|\sup |\alpha_i \xi_i|\|_{p,\infty}^p = \sup_{\lambda > 0} \lambda^p \mathbb{P}(\sup |\alpha_i \xi_i| > \lambda) \leq \sup_{\lambda > 0} \lambda^p \sum_i \mathbb{P}(|\alpha_i \xi_i| > \lambda)$$

and second, $\lim \alpha_i \xi_i = 0$ a.s. since the series $\sum \mathbb{P}(|\alpha_i, \xi_i| > \lambda)$ converges for any $\lambda > 0$.

1) $\Rightarrow$ 4). By Lemma 2.2, there exists $\varepsilon_0 > 0$ such that the series $\sum \mathbb{P}(|\alpha_i \xi| > 1)$ converges for all $\xi \in L_{p,q}$, $\|\xi\| \leq \varepsilon_0$. This implies, in turn, that the series $\sum_{j=1}^{\infty} \rho_{-j}(a)\rho_j(\xi)$ also converges. Put

$$\delta = \min\{[m(p,q)\varepsilon_0]^p, 2^p - 1\}$$

(where $m(p,q)$ is as in Lemma 1.1) and let $\{\gamma_j\}$ be an arbitrary sequence of nonnegative numbers such that $\|\{\gamma_j\}\|_{q/p} \leq \delta$. Define a r.v. ξ_0 on Ω by

$$\rho_j(\xi_0) = \begin{cases} 0, & \text{if } j \leq 0, \\ 2^{-jp}\gamma_j, & \text{if } j > 0. \end{cases}$$

Using Lemma 1.1, one easily verifies that $\|\xi_0\|_{p,q} \leq \varepsilon_0$; therefore, the series $\sum_{j=1}^{\infty} \rho_{-j}(a)\rho_j(\xi_0)$, which can also be written as $\sum_{j=1}^{\infty} 2^{-jp}\rho_{-j}(a)\gamma_j$ converges. Since $\{\gamma_j\}$ is an arbitrary nonnegative sequence such that $\|\{\gamma_j\}\|_{q/p} \leq \delta$, we have $(2^{-jp}\rho_{-j}(a)) \in \ell_{r/p}$, where $r^{-1} + q^{-1} = p^{-1}$ if $p < q$, and $r = \infty$ if $p \geq q$. But this means that $a \in \ell_{p/r}$, completing the proof.

Thus,

$$A^{\infty}(p,q) = \begin{cases} \ell_{p,r}, & r^{-1} + q^{-1} = p^{-1}, \quad \text{if } 0 < p \leq \infty; \\ \ell_{p,\infty}, & \text{if } 0 < q \leq p < \infty. \end{cases}$$

§3. The classes $A^0(p,q)$

In this section we study almost sure convergence of series of independent r.v.'s. Our investigation will naturally rely on Kolmogorov's 3-series test:

THEOREM 3.1 (Kolmogorov, 1929). *Let $\{\xi_i\}$ be a sequence of independent r.v.'s. The series $\sum \xi_i$ converges if and only if the following three conditions are satisfied*:

1) $\sum \mathbb{P}(|\xi_i| > 1) < \infty$;
2) $\sum \mathbb{E}\xi_i'$ *converges*;
3) $\sum \mathbb{D}\xi_i' < \infty$, *where* $\xi_i' = \xi_i \cdot 1_{[|\xi_i| \leq 1]}$.

Here $\mathbb{E}$ and $\mathbb{D}$ denote expectation and variance, respectively.

Remark that condition 1 implies $\sup|\xi_i| < \infty$ a.s. Condition 2 holds for any sequence of symmetric r.v.'s.

We now state the main results of this paper.

THEOREM 3.2. *If $0 < p < 2$ and $0 < q \leq \infty$, then*

$$A^0(p,q) = A^{\infty}(p,q).$$

PROOF. The proof for $0 < q \leq p$ follows easily from the Marcinkiewicz-Zygmund theorem on random series. The reduction was carried out in [5], where it was shown that $A^0(p,q) = \ell_{p,\infty}$, $0 < p < 2$. Now, taking the embedding $L_{p,q} \subset L_p$ into account, we obtain $A^0(p,p) \subset A^0(p,q) \subset A^\infty_{p,q} = \ell_{p,\infty}$. Thus, $A^0(p,q) = A^\infty(p,q) = \ell_{p,\infty}$, $0 < p < 2$, $0 < q \le p$.

We now proceed to the case $0 < p < q < \infty$. It is obvious that $A^0(p,q) \subset A^\infty(p,q) = \ell_{p,r}$, $r = pq/(q-p)$. Let $a = \{\alpha_i\} \in \ell_{p,r}$; $\{\xi_i\} \in \Phi(p,q)$. Conditions 1 and 2 of Theorem 1 hold for the series $\sum \alpha_i \xi_i$. In addition we may assume without loss of generality that $1 \ge |\alpha_1| \ge |\alpha_2| \ge \cdots$. We now verify Condition 3 of Theorem 1:

$$\sum_{i=1}^\infty \mathbb{E}[(\alpha_i \xi_i)']^2$$

$$\le \sum_{k=1}^\infty 2^{-2k} \rho_{-k}(a) \mathbb{E}(|\xi_1|^2 \cdot 1_{[|\xi_1| \le 2^k]})$$

$$= \sum_{j=0}^\infty \sum_{k=j}^\infty 2^{-2k} \rho_{-k}(a) \int_{A_j} |\xi_1|^2 d\mathbb{P}$$

(here $A_j = \{2^j < |\xi_1| \le 2^{j+1}\}$, $j = 0, \pm 1, \ldots$)

$$= 4 \sum_{j=0}^\infty 2^{pj} \rho_j(\xi_1) \cdot 2^{(2-p)j} \sum_{k=j}^\infty 2^{-2k} \rho_{-k}(a)$$

$$\le 4 \left(\sum_{j=0}^\infty 2^{qj} \rho_j^{q/p}(\xi_1) \right)^{p/q} \left(\sum_{j=0}^\infty 2^{(2-p)j \frac{q}{q-p}} \left(\sum_{k=j}^\infty 2^{-2k} \rho_{-k}(a) \right)^{\frac{q}{q-p}} \right)^{\frac{q-p}{q}},$$

by the Hölder inequality with exponents q/p and $(q/p)'$. The first factor in the last expression is finite by Lemma 1.1 To estimate the second factor, we apply one of the numerous variants of the Hardy inequality: If $d \ge 1$, $\delta > 0$, then

$$\sum_{i=0}^\infty 2^{i\delta} \left(\sum_{j=i}^\infty a_j \right)^d \le C \sum_{i=0}^\infty 2^{i\delta} a_i^d,$$

where $c > 0$ is a constant (for a proof of this inequality see, e.g., [4]). By Lemma 1.1,

$$\sum_{j=0}^\infty 2^{(\frac{2-p}{q-p})qj} \left(\sum_{k=j}^\infty 2^{-2k} \rho_{-k}(a) \right)^{\frac{q}{q-p}}$$

$$= \sum_{j=0}^\infty 2^{jr(\frac{2}{p}-1)} \left(\sum_{k=j}^\infty 2^{-2k} \rho_{-k}(a) \right)^{\frac{r}{p}}$$

$$\le C \sum_{j=0}^\infty 2^{ir(\frac{2}{p}-1)} (2^{-2j} \rho_{-j}(a))^{\frac{r}{p}}$$

$$= C \sum_{j=0}^\infty 2^{-jr} (\rho_{-j}(a))^{\frac{r}{p}} < \infty.$$

Thus, the third condition of Theorem 1 is satisfied, and the series $\sum \alpha_i \xi_i$ converges a.s., i.e., $a \in A^0(p, q)$.

To complete the proof, it remains to consider the case $0 < p < 2$, $q = \infty$. The above estimates may be carried out with obvious modifications, but it is easier to obtain the result directly. If $\xi_1 \in L_{p,\infty}$, then $\mathbb{P}(|\xi_1| > t) \le k t^{-p}$, $t > 0$. Therefore, $\mathbb{E}[(\alpha_i \xi_i)']^2 \le |\alpha_i|^2 + (2/(2-p))(|\alpha_i|^p - |\alpha_i|^2)$. By this inequality, $\sum \mathbb{E}[(\alpha_i \xi_i)']^2 < \infty$ for $a = \{\alpha_i\} \in \ell_p$. On the other hand, $A^\infty(p, \infty) = \ell_p$, $0 < p < \infty$. This completes the proof.

THEOREM 3.3. *If either* $p = 2$, $0 < q \le 2$, *or* $p > 2$, $0 < q \le \infty$, *then* $A^0(p, q) = \ell_2$.

The proof is very simple. On one hand, the embedding $L_{p,q} \subset L_2$ yields $A^0(2, 2) \subset A^0(p, q)$, i.e., $\ell_2 \subset A^0(p, q)$. On the other hand, considering a standard sequence of Bernoulli r.v.'s with $\mathbb{P}(\varepsilon_i = 1) = \mathbb{P}(\varepsilon_i = -1) = 1/2$, we obtain $A^0(p, q) \subset \ell_2$.

In the next theorem we shall make our first excursion from the spaces $L_{p,q}$ and become acquainted with (or recall) another important class of sequence spaces. A nondecreasing function $M : [0, \infty) \to [0, \infty)$ that is continuous at zero and satisfies the conditions $M(0) = 0$, and $M(t) > 0$ for $t > 0$, is called an *Orlicz function*. The Orlicz space ℓ_M is the vector space of all real sequences $x = \{x_n\}$ such that the series $\sum M(|\varepsilon x_n|)$ is convergent for some $\varepsilon > 0$. Defining $B_M(\delta) = \{x : \sum F(|x_n|) < \delta\}$ we see that the sets $\{rB_M(\delta) : r > 0, \delta > 0\}$ form a base for the topology of the complete metrizable linear space ℓ_M. If M is a convex function, ℓ_M is a normed space. If M satisfies the so-called Δ_2-condition at zero, i.e.,

$$\sup_{0 < t \le 1} M(2t)/M(t) < \infty,$$

then $\ell_M = \{x : \sum M(|x_n|) < \infty\}$. It is clear that the definition of ℓ_M depends only on the behavior of M in the neighborhood of 0. Two Orlicz functions M and N are said to be equivalent if

$$0 < \inf_{0 < t \le 1} \frac{M(t)}{N(t)} \le \sup_{0 < t \le 1} \frac{M(t)}{N(t)} < \infty.$$

The Orlicz spaces corresponding to equivalent functions coincide as sets and their topologies are equivalent.

We will consider the Orlicz spaces corresponding to the following four functions:

$$\begin{aligned} M_0(t) &= t^2 \ln(1 + t^{-1}), \\ M_1(t) &= t^2 |\ln t|, \\ M_2(t) &= t^2 \ln(1 + t^{-2}), \\ N(t) &= t \ln(1 + t^{-1}), \qquad t > 0. \end{aligned}$$

It is easy to verify that the functions $M_i(t)$, $i = 0, 1, 2$, are equivalent at zero.

THEOREM 3.4. *We have* $A^0(2, \infty) = \ell_{M_0}$.

PROOF. It follows from the estimate
$$\mathbb{E}[(\alpha_i\xi_i)']^2 = 2\alpha_i^2 \int_0^{2/|\alpha_i|} t\mathbb{P}(|\xi_i| > t)\,dt \leq \alpha_i^2(1 + 2|\ln|\alpha_i||),$$

Theorem 3.1, the equality $A^\infty(2, \infty) = \ell_2$, and the embedding $\ell_{M_0} \subset \ell_2$ that $\ell_{M_0} \subset A^0(2, \infty)$. To prove the converse we consider the sequence $\{\xi_i\} \in \Phi(2, \infty)$ with the Pareto distribution:
$$\mathbb{P}(|\xi_1| > t) = \begin{cases} t^{-2}, & t \geq 1, \\ 1, & 1 > t \geq 0. \end{cases}$$

This completes the proof.

We have not yet considered the case $p = 2$, $2 < q < \infty$; here we cannot give an exact answer, but only prove the following result.

PROPOSITION. *Let* $a = \{\alpha_i\}$ *be a monotone decreasing sequence,* $a \in \ell_{2,r} = A^\infty(2, q)$,
$$\sum_{j=1}^\infty \left[(\alpha_{j-1}/\alpha_j)^2 \sum_{n=j}^\infty \alpha_n^2\right]^{q/(q-2)} < \infty.$$

Then $a \in A^0(2, q)$.

§4. The classes $A(p, q)$

In this section we consider the normed $L_{p,q}$ spaces. Before proving the main results, we must do some preliminary work.

A r.v. $f^{(r)}$ is said to be *r-stable* $(0 < r \leq 2)$ if $\mathbb{E}\exp(itf^{(r)}) = \exp(-c|t|^r)$ for some constant $c > 0$ and all $t \in \mathbb{R}$. Let $\{f_i^{(r)}\}_{i=1}^\infty$ be a sequence of independent r-stable r.v.'s, $0 < r < 2$, and $K(\{f_i^{(r)}\})$ be the closed linear span of this sequence ("closed" with respect to the probability). Then $K(\{f_i^{(r)}\}) \in L_{r,\infty}$, and
$$b\left(\sum_{i=1}^n |\alpha_i|^r\right)^{1/r} \leq \left\|\sum_{i=1}^n \alpha_i f_i^{(r)}\right\|_{L_{r,\infty}} \leq B\left(\sum_{i=1}^n |\alpha_i|^r\right)^{1/r}.$$

Thus, it is possible to construct a "through" subspace, isomorphic to ℓ_r for the following sequence of spaces of r.v.'s:
$$L_{r,\infty} \subset L_{p,1} \subset L_p \subset L_{p,q} \subset L_{p,\infty} \subset L_1 \subset L_0,$$
where $1 < p < r < 2$, $p < q < \infty$.

The conditional expectation operator $\mathbb{E}^{\mathscr{F}_1}$ with respect to the σ-algebra $\mathscr{F}_1 \subset \mathscr{F}$ has norm 1 as an operator in L_1 and L_∞. Consequently, it acts in any (normed) space $L_{p,q}$. The following inequality follows from this observation: if X and Y are independent r.v.'s such that $\mathbb{E}X = 0$, then $\|X+Y\|_{p,q} \geq \|Y\|_{p,q}$, $1 \leq p < \infty$, $1 \leq q < \infty$.

LEMMA 4.1. *Let* $1 \leq p < \infty$, $1 \leq q < \infty$. *If* $\{\alpha_i\} \in A(p,q)$, *then there exists* $C = C(\{\alpha_i\}) > 0$ *such that*

$$\left\| \sum_{i=1}^{\infty} \alpha_i f_i \right\|_{p,q} \leq c \|f_1\|_{p,q},$$

where $\{f_i\}$ *is a sequence of independent identically distributed r.v.'s,* $\mathbb{E}f_1 = 0$.

PROOF. Let $L^0_{p,q} = \{f \in L_{p,q} : \mathbb{E}f = 0\}$. Using the symmetrization one can prove that if $\{\alpha_i\} \in A(p,q)$, then $\sum \alpha_i f_i$ converges in $L_{p,q}$ for any sequence $\{f_i\}$ of independent identically distributed r.v.'s such that $\mathbb{E}f_1 = 0$. Define a new norm on $L^0_{p,q}$ by

$$\|f\|^{(1)} = \left\| \sum_{i=1}^{\infty} \alpha_i f_i \right\|_{p,q}, \qquad f_i \stackrel{d}{=} f.$$

The equivalence of two norms on $L^0_{p,q}$ can be proved by standard methods based on the Closed Graph Theorem.

LEMMA 4.2. *Let* $1 \leq p < 2$, $1 \leq q < \infty$. *If* $a = \{\alpha_k\} \in A(p,q)$, *then* $\sup_{r>p} \sum_{k=1}^{\infty} |\alpha_k|^r < \infty$.

PROOF. We first observe that $\sum |\alpha_k|^r < \infty$ for any $r > p$, since $\Phi(p,q)$ contains a sequence of r-stable r.v.'s. Suppose that $\sup_{r>p} \sum_{k=1}^{\infty} |\alpha_k|^r = \infty$ and that there is a sequence $2 \geq r_n \searrow p$ such that

$$\sum_{k=1}^{\infty} |\alpha_k|^{r_n} > 2^{3n}, \qquad n = 1, 2, \ldots.$$

We construct a double sequence $\{Y_{nk}\}_{n,k=1}^{\infty}$ of independent r.v.'s as follows:
  a) for each fixed n, Y_{nk}, $k = 1, 2, \ldots$ is a sequence of identically distributed r_n-stable r.v.'s;
  b) $\|Y_{nk}\|_{p,q} = 1$; $n = 1, 2, \ldots$; $k = 1, 2, \ldots$.

This can be done by suitably choosing the normalizing constant C in the definition of an r-stable r.v. When this is done,

$$\left\| \sum_{k=1}^{m} \alpha_k Y_{nk} \right\| = \left[\sum_{k=1}^{m} |\alpha_k|^{r_n} \right]^{1/r_n}.$$

Let $Z_k := \sum_{n=1}^{\infty} Y_{nk}/2^n$ (convergence in $L_{p,q}$), $k = 1, 2, \ldots$. The sequence $\{Z_k\}$ is formed by independent symmetric identically distributed r.v.'s, i.e.,

$\{Z_k\} \in \Phi(p, q)$. Fixing $m \in \mathbb{N}$, consider

$$\left\| \sum_{k=1}^{m} \alpha_k Z_k \right\|_{p,q} = \left\| \sum_{k=1}^{m} \alpha_k \sum_{n=1}^{\infty} Y_{nk}/2^n \right\| = \left\| \sum_{n=1}^{\infty} \sum_{k=1}^{m} \alpha_k Y_{nk}/2^n \right\|$$
$$\geq \left\| \sum_{k=1}^{m} \alpha_k Y_{nk}/2^n \right\| = \left[\sum_{k=1}^{m} |\alpha_k|^{r_n}/2^{nr_n} \right]^{1/r_n} = \frac{1}{2^n} \left[\sum_{k=1}^{m} |\alpha_k|^{r_n} \right]^{1/r_n},$$
$$n = 1, 2, \ldots.$$

This chain of inequalities makes use of the opening remark of this section. We now have

$$\left\| \sum_{k=1}^{\infty} \alpha_k Z_k \right\|_{p,q} \geq \frac{1}{2^n} \left[\sum_{k=1}^{\infty} |\alpha_k|^{r_n} \right]^{1/r_n} > \frac{1}{2^n} \cdot 2^{3n/r_n}$$
$$= 2^{n(3/r_n - 1)} > 2^{n(3/2 - 1)} = 2^{n/2}, \quad n = 1, 2, \ldots.$$

This contradicts Lemma 1, proving Lemma 4.2.

THEOREM 4.3. *Let $1 \leq p < 2$, $1 \leq q < \infty$. If $a = \{\alpha_k\} \in A(p, q)$, then $a \in \ell_p$.*

PROOF. First let $q < \infty$. Denote $M := \sup_{r>p} \sum_{k=1}^{\infty} |\alpha_k|^r < \infty$ and consider a sequence $r_n \searrow p$. For fixed m, we have

$$\sum_{k=1}^{m} |\alpha_k|^{r_n} \leq M, \quad n = 1, 2, \ldots.$$

Therefore, $\sum_{k=1}^{m} |\alpha_k|^p \leq M$, $m \in \mathbb{N}$, i.e., $a = \{\alpha_k\} \in \ell_p$. If $q = \infty$, the result follows immediately from the fact that $A(p, \infty) \subset A^0(p, \infty) = \ell_p$. This proves the theorem.

The following inequality was proved in [4]:

$$\left\| \sum_{i=1}^{n} \xi_i \right\|_{p,q}^{p} \leq C \sum_{i=1}^{n} \|\xi_i\|_{p,q}^{p},$$

where $\{\xi_i\}$ are independent symmetric r.v.'s. Combining this inequality with Theorem 4.3, we obtain $A(p, q) = \ell_p$, $1 \leq p < 2$, $p \leq q \leq \infty$. It is interesting that for these values of p and q the families of subspaces generated by random variables for different q are quite different. While it is not difficult to find a sequence of independent identically distributed r.v.'s in $L_{p,\infty}$ that is equivalent to the standard basis in ℓ_p (for example, the p-stable random variables), no such sequence exists in L_p. Indeed, the following proposition can be proved:

PROPOSITION. *If $1 \leq p < 2$, $\{\xi_k\} \in \Phi(p, p)$, then*

$$\lim_{n \to \infty} \frac{\| \sum_{k=1}^{n} \xi_k \|_p}{n^{1/p}} = 0.$$

A proof of this proposition may be derived from the Berry-Esseen inequality.

We now impose the following restrictions on the indices: $1 \leq p < 2$, $1 \leq q < p$. Then we can prove the following analog of Theorem 4.3.

THEOREM 4.4. *Let $1 \leq p < 2$, $1 \leq q < p$. If $a = \{\alpha_k\} \in A(p, q)$, then $a \in \ell_{p,q}$.*

PROOF. Let $a \in A\{p, q\}$ and $\{\xi_i\} \in \Phi(p, q)$. By Lévy's inequality and Lemma 1, there exists a constant $c > 0$ such that

$$\| \sup_{1 \leq i < \infty} |\alpha_i \xi_i| \|_{p,q} \leq c.$$

Let $\{\gamma_i\}$ be an arbitrary sequence of nonnegative numbers such that $\|\{\gamma_i\}\|_{q/p} \leq \delta$, where δ is a positive number. We construct a r.v. ξ as in the proof of the implication 1) $\Rightarrow$ 4) in Theorem 2.3,

$$p_j(\xi) = \begin{cases} 0, & \text{if } j \leq 0, \\ 2^{-jp}, & \text{if } j > 0. \end{cases}$$

This is possible if δ is sufficiently small. With this construction, $\xi \in L_{p,q}$ and $\|\xi\|_{p,q} \leq \varepsilon(\delta)$, where $\varepsilon(\delta) \to 0$ as $\delta \to +0$; $\mathbb{P}(|\xi| > 0) \leq \lambda(\delta)$, $\lambda(\delta) \to 0$ as $\delta \to +0$. Thus, choosing δ sufficiently small and using the independence condition, we can assume that

$$\sigma_j \left(\sup_{1 \leq i \leq n} |\alpha_i \xi_i| \right) = \mathbb{P} \left(\sup_{1 \leq i \leq n} |\alpha_i \xi_i| > 2^j \right)$$

$$\geq \mathbb{P}(|\alpha_1 \xi_1| > 2^j) + \sum_{i=2}^{n} \mathbb{P} \left(|\alpha_i \xi_i| > 2^j, \sup_{1 \leq k \leq i-1} |\alpha_k \xi_k| \leq 2^j \right)$$

$$= \mathbb{P}(|\alpha_1 \xi_1| > 2^j) + \sum_{i=2}^{n} \mathbb{P}(|\alpha_i \xi_i| > 2^j) \prod_{1 \leq k \leq i-1} \mathbb{P}(|\alpha_k \xi_k| \leq 2^j)$$

$$\geq c \sum_{i=1}^{n} \sigma_j(\alpha_i \xi_i), \quad \xi_i \stackrel{d}{=} \xi, \quad i = 1, 2, \ldots.$$

Therefore, the operator $\{\gamma_j\} \to \{\sum_{k=1}^{\infty} \rho_{-k}(a) 2^{-kp} \gamma_{j+k}\}$ is bounded in $\ell_{q/p}$. Hence $\sum_{k=1}^{\infty} (\rho_{-k}(a) 2^{-kp})^{q/p} < \infty$, i.e., $a \in \ell_{p,q}$. This completes the proof.

We have proved that $A(p, q) \subset \ell_{p,q}$ for $1 \leq p \leq 2$, $1 \leq q \leq p$. As A. M. Shteinberg has proved the reverse inclusion, we have $A(p, q) = \ell_{p,q}$, $1 \leq p \leq 2$, $1 \leq q \leq p$. Since the proof of Theorem 4.4 does not use the inequality $q \leq p$, the theorem is true for $q > p$ as well; in this case, however, the result is no longer sharp (cf. Theorems 4.3 and 4.4). All we know is that $A(2, 2) = \ell_2$, $A(2, \infty) = A^0(2, \infty)$. The first equality is trivial. We now prove the second.

THEOREM 4.5. $A^0(2, \infty) \subset A(2, \infty)$.

LEMMA. *Let $\{\alpha_j\}$ be a sequence of positive numbers. If $\sum_j \alpha_j \ln(1+1/\alpha_j) < \infty$, then the series $\sum_j \alpha_j e_j$ is convergent in $L_{1,\infty}$ for any sequence $\{e_j\}$ such that $e_j \geq 0$, a.s., and $\mathbb{P}(e_i > t) = 1/t$, $t \geq 1$, $i = 1, 2, \ldots$.*

PROOF. The function $\psi(u) = u\ln(1+1/u)$, $u > 0$, is concave, increasing, and satisfies the condition $\psi(u) \to 0$ as $u \to +0$, $\psi(u) \to 1$ as $u \to \infty$. Fix $a > 0$ and a sequence of positive numbers $\alpha_1, \alpha_2, \ldots$ such that $\sum_j \psi(\alpha_j) \leq a$. Then $\sum_j \alpha_j \leq \psi^{-1}(a)$. Let $e_j \geq 0$ a.s. and $\mathbb{P}(e_j > t) = 1/t$, $t \geq 1$, $j = 1, 2, \ldots$. If $\Delta \in \mathscr{F}$ and $\mathbb{P}(\Delta) = t > 0$, then

$$t \inf_{\omega \in \Delta} \sum_{j=1}^n \alpha_j e_j(\omega) \leq \int_\Delta \left(\sum_{j=1}^n \alpha_j e_j\right) d\mathbb{P}.$$

Denote

$$\Delta_j := \left\{\omega \in \Delta : e_j(\omega) > \frac{2\psi^{-1}(a)}{\alpha_j t}\right\}.$$

Since

$$\mathbb{P}(\Delta_j) \leq \mathbb{P}\left(e_j > \frac{2\psi^{-1}(a)}{\alpha_j t}\right) = \frac{\alpha_j t}{2\psi^{-1}(a)}, \qquad j = 1, 2, \ldots,$$

it follows that

$$\mathbb{P}\left(\Delta \setminus \bigcup_{j=1}^n \Delta_j\right) \geq t - \frac{t}{2\psi^{-1}(a)} \sum_{j=1}^n \alpha_j \geq t - \frac{t}{2} = \frac{t}{2}, \qquad n = 1, 2, \ldots.$$

Now we have

$$\inf_\Delta \sum_{j=1}^n \alpha_j e_j \leq \inf_{\Delta \setminus \cup_{j=1}^n \Delta_j} \sum_{j=1}^n \alpha_j e_j \leq \frac{1}{\mathbb{P}(\Delta \setminus \cup_{j=1}^n \Delta_j)} \int_{\Delta \setminus \cup_{j=1}^n \Delta_j} \left(\sum_{j=1}^n \alpha_j e_j\right) d\mathbb{P}$$

$$\leq \frac{2}{t} \sum_{j=1}^n \alpha_j \int_{\Delta \setminus \cup_{j=1}^n \Delta_j} e_j \, d\mathbb{P} \leq \frac{2}{t} \sum_{j=1}^n \alpha_j \int_{\{\omega \in \Delta : e_j \leq 2\psi^{-1}(a)/(\alpha_j t)\}} e_j \, d\mathbb{P}$$

$$\leq \frac{2}{t} \sum_{j=1}^n \alpha_j \int_{\alpha_j t/2\psi^{-1}(a)}^{\alpha_j t/2\psi^{-1}(a)+t} e_j^*(s) \, ds = \frac{2}{t} \sum_{j=1}^n \alpha_j \int_{\alpha_j t/2\psi^{-1}(a)}^{\alpha_j t/2\psi^{-1}(a)+t} \frac{ds}{s}$$

$$= \frac{2}{t} \sum_{j=1}^n \alpha_j \ln(1 + 2\psi^{-1}(a)/\alpha_j)$$

$$\leq \frac{2}{t}(1 + 2\psi^{-1}(a)) \sum_{j=1}^n \alpha_j \ln\left(1 + \frac{1}{\alpha_j}\right)$$

$$\leq \frac{2a}{t}(1 + 2\psi^{-1}(a)).$$

Therefore,
$$\inf_{\Delta} \sum_{j=1}^{n} \alpha_j e_j \leq \frac{2a}{t}(1 + 2\psi^{-1}(a)),$$

and $t(\sum \alpha_j e_j)^*(t) \leq 2a(1 + 2\psi^{-1}(a))$. This proves the lemma.

PROOF OF THE THEOREM. Let $b = \{\beta_j\} \in A_0(2, \infty) = \ell_{M_0}$, where $M_0(u) = u^2 \ln(1 + 1/u)$, $u > 0$. Since the functions $M_0(u)$ and $M_2(u) = u^2 \ln(1 + u^{-2})$ are equivalent at zero,

$$\sum_{i=1}^{\infty} \beta_i^2 \ln(1 + \beta_i^{-2}) < \infty,$$

i.e., $\{\beta_i^2\}$ satisfies the assumptions of the lemma. Let $\{\xi_i\}$ be a sequence of independent identically distributed symmetric r.v.'s with the "extreme" distributions for $L_{2,\infty}$ i.e.,

$$\mathbb{P}(|\xi_i| > t) = t^{-2}, \qquad t \geq 1, \ i = 1, 2, \ldots.$$

Observe that $\mathbb{P}(\xi_i^2 > t^2) = t^{-2}$ or $\mathbb{P}(\xi_i^2 > s) = s^{-1}$, $s \geq 1$. By the lemma, the series $\sum \beta_i^2 \xi_i^2$ converges in $L_{1,\infty}$. But

$$\|\sum \beta_i \xi_i\|_{2,\infty}^2 \asymp \|(\sum |\beta_i \xi_i|^2)^{1/2}\|_{2,\infty}^2 = \|\sum \beta_i^2 \xi_i^2\|_{1,\infty}.$$

Finally, if $\{\eta_i\}$ is an arbitrary sequence in $\Phi(2, \infty)$, then $\mathbb{P}(|\eta_i| > t) \leq ct^{-2} = c\mathbb{P}(|\xi_i| > t)$, $t \geq 1$, and we can use the comparison theorem [2, Chapter V, Theorem 4.4]. Hence the series $\sum \beta_i \eta_i$ is convergent in $L_{2,\infty}$. This means that $b \in A(2, \infty)$, proving the theorem.

It is interesting to compare Theorem 4.5 with the following result of Marcus and Pisier (see [6]): for any sequence of positive independent r.v.'s $\{z_n\}$ there exists $\lambda > 0$ such that

$$\|\|\{z_n\}\|_{p,\infty}\|_{p,\infty} \leq \lambda \left(\sum_n \|z_n\|_{p,\infty}^p\right)^{1/p}$$

for $0 < p < \infty$. Here $\|\{z_n\}\|_{p,\infty} = \sup_n n^{1/p}\{z_n\}^*$ is the norm in the sequence space. Using this inequality, one immediately proves the implication

$$b = \{\beta_i\} \in l_2 \Rightarrow \|\{\beta_i \eta_i\}\|_{2,\infty} \in L_{2,\infty}.$$

Theorem 4.5 can be expressed in similar notation.

We conclude with a few words about the equality $A(p, q) = \ell_2$, where $p > 2$, $1 < q < \infty$. The inclusion $A(p, q) \subset \ell_2$ can be derived using a sequence of Bernoulli r.v.'s. The inclusion $\ell_2 \subset A(p, q)$ is easily explained for $p \leq q < \infty$, since in that case $L_{p,q}$ is 2-convex, i.e., there exists $c > 0$ such that

$$\|(\sum |X_k|^2)^{1/2}\| \leq C(\sum \|X_k\|^2)^{1/2}$$

for any finite collection of elements $X_1, X_2, \ldots, X_n$ in $L_{p,q}$.

References

1. M. Stein and G. Weiss, *Introduction to Fourier analysis on Euclidean spaces*, Princeton Univ. Press, Princeton, N. J., 1971.
2. N. N. Vakhaniya, V. I. Tarieladze, and S. A. Chobanyan, *Probability distributions on Banach spaces*, "Nauka", Moscow, 1985; English transl., Reidel, Dordrecht, 1987.
3. J. Lindenstruss and L. Tzafriri, *Classical Banach spaces*, Springer-Verlag, Berlin, 1979.
4. R. Norvaisha and A. Rachkauskas, *The law of large numbers with respect to quasinorms*, Litovsk. Mat. Sb. **24** (1984), no. 2, 130–144; English transl. in Lithuanian Math. J. **24** (1984).
5. R. Ulbricht, *Weighted sums of independent identically distributed random variables*, Ann. Probab. **9** (1981), 633–638.
6. M. Marcus and G. Pisier, *Characterizations of almost surely continuous p-stable random Fourier series and strongly stationary processes*, Acta Math. **152** (1984), 245–301.
7. J. Rosinski, *Remarks on Banach spaces of stable type*, Probab. Math. Statist. **1** (1980), 67–71.

Translated by V. OPERSTEIN

Corner Singularities and Multidimensional Folds in Nonlinear Analysis

YU. I. SAPRONOV

Corner singularities are generalizations of boundary singularities defined by V. I. Arnol'd [1]. They were apparently first introduced and studied by D. Siersma [2]. Due to the natural connection between corner singularities and singularities of functions that are even in each of the variables, one can apply the theory of equivariant singularities [3, 4]. In nonlinear analysis corner singularities arise in variational problems with *semiconstraints* and in variational problems that are equivariant under systems of involutions commuting on the linear spans of fundamental bifurcation modes.

§1. Finite-dimensional reductions

A smooth functional V on a smooth Banach manifold M, modeled by a space E, is said to have a finite-dimensional singularity type at a critical point a if there exists a diffeomorphism $\varphi : O^n \times O^{\infty-n} \to U$, $\varphi(0, 0) = a$, such that the origin is the only critical point of $V(\varphi(\xi, \cdot))$ in $O^{\infty-n}$ for any $\xi \in O^n$, and moreover it is a weakly regular critical point (the second differential is an algebraically nondegenerate quadratic form). Here U, O^n, and $O^{\infty-n}$ are neighborhoods in M, $\mathbb{R}^n$, and $E^{\infty-n}$, where the latter is a linear subspace of codimension n in E. In applications the finite dimensionality of the type of a singularity is very important if one uses the Lyapunov-Schmidt method [5–7]. If the value of $V(\varphi(\xi, \cdot))$ at the origin is a local minimum for any $\xi \in \mathbb{R}^n$, then the behavior of V near a is completely determined by the (key) function $W(\xi) = V(\varphi(\xi, 0))$. If a smooth mapping $g : U \to \mathbb{R}^m$, $g = (g_1, \ldots, g_m)$ such that $g(\varphi(\xi, u)) = \xi$, is defined on U, then φ is said to be compatible with g. If $V|_{M_\xi}$ achieves the minimum at $\varphi(\xi, 0)$, $\forall \xi$ (where $M_\xi = U \cap g^{-1}(M)$), the key function is defined as $W(\xi) = \inf_{x \in M_\xi} V(x)$. The reduction of V to W also replaces the analysis of V on the m-faced corner $\{x \in U | g_k(x) \geq 0\}_{k=1}^{m}$ by analysis of W on $O^n \cap (\mathbb{R}_+^m \times \mathbb{R}^{n-m})$, where $\mathbb{R}_+^m = \{\eta \in \mathbb{R}^m | \eta_j \geq 0\}$ is a simplicial cone. If a Lie group G acts smoothly on U and $\mathbb{R}^n$ in such a way that V is invariant and g equivariant with respect to the action of G, the key function inherits the property of G-invariance.

§2. Semiregular corner singularities

An n-dimensional manifold with an m-faced corner ($m \leq n$) is a smooth real n-dimensional manifold with a collection of m nonsingularly intersecting smooth hypersurfaces. Locally (in a neighborhood of the corner point) such a manifold is similar to $\mathbb{R}^n$ with a distinguished family of hyperplanes $\mathbb{R}_j^n = \{x \in \mathbb{R}^n | x_j = 0\}$, $1 \leq j \leq m$ (where x_j is a coordinate of the point x). If $m = 1$ we get the notion of a manifold with boundary [1]. The set $C = \{x \in \mathbb{R}^n | x_j \geq 0, \forall j \leq m\}$ is called an m-faced corner. A point $a \in C$ is said to be a *conditionally critical* (c.c.) point of a smooth function W in $\mathbb{R}^n$ if $\mathrm{grad}\, W(a)$ is orthogonal to the face of C that contains a. The number of elements in the support $\mathrm{supp}(a) = \{k \in \{1, \ldots, m\} | x_k \neq 0\}$ is called the order of a. The multiplicity $\hat{\mu}$ of a c.c. point $a \in C$ is defined as the dimension of the quotient algebra $Q = \mathbb{R}[[x-a]]/I(W, a)$, where $\mathbb{R}[[x-a]]$ is the algebra of formal power series in $x - a$ and $I(W, a)$ the ideal in $\mathbb{R}[[x-a]]$ generated by the sets $\{\frac{\partial W}{\partial x_j}\}_{j \notin K}$ and $\{x_k \frac{\partial W}{\partial x_k}\}_{k \in K}$, where $K = \{1, \ldots, m\} \setminus \mathrm{supp}(a)$. A c.c. point is called regular if $\hat{\mu} = 1$. Let $\mu_k(W, a)$ denote the (usual) multiplicity of the restriction $V|_{\mathbb{R}_k^n}$ at a, where $\mathbb{R}_K^n = \bigcap_{k \in K} \mathbb{R}_k^n$, $K \subset \{1, \ldots, m\} \setminus \mathrm{supp}(a)$. A c.c. point is called semiregular if $\mu_k \leq 1$, $\forall K \subset \{1, \ldots, m\} \setminus \mathrm{supp}(a)$. If $\mathrm{supp}(a) = \varnothing$ and $\mu_k = 1$, for any k, then there is a corner change of coordinates (i.e., a diffeomorphism that maps each face of C into itself) that reduces W to normal form:

$$\sum_{k=1}^{m} \sigma_k x_k^2 + \sum_{1 \leq k_1 < \cdots < k_q \leq m, q \geq 2} a_{k_1, \ldots, k_q} x_{k_1} \cdots x_{k_q} + \sum_{p=1}^{n} x_p^2, \qquad (1)$$

where $\sigma_k \in \{-1, 1\}$. A canonical miniversal deformation of a singularity of W in the form (1) at the origin is given by the evolute

$$W(x) + \sum_{1 \leq k_1 < \cdots < k_q \leq m} \lambda_{k_1, \ldots, k_q} x_{k_1} \cdots x_{k_q}, \qquad \lambda = \{\lambda_{k_1, \ldots, k_q}\}. \qquad (2)$$

Proofs of these statements follow from their equivariant analogues [1–4].

For a singularity (1) we may assume that $m = n$. If $\widehat{\mathbb{R}}^n$ is a space with coordinates $y_1, \ldots, y_n$, which is 2^n-fold covered by the map $\pi : y \to x = (y_1^2, \ldots, y_k^2)$, then W can be lifted to $\widehat{\mathbb{R}}^n$ by the formula $\widehat{W}(y) = W(\pi(y))$. The function W is invariant under involutions $J_1, \ldots, J_n$, where J_k changes the sign of the kth coordinate. To each corner diffeomorphism $\varphi : (\mathbb{R}^n, 0) \to (\mathbb{R}^n, 0)$ there corresponds a diffeomorphism $\widehat{\varphi} : (\widehat{\mathbb{R}}^n, 0) \to$

($\widehat{\mathbb{R}}^n$, 0) equivariant under $J_1, \ldots, J_n$, such that $\varphi \cdot \pi = \pi \cdot \widehat{\varphi}$ and $\widehat{\varphi}(\widehat{C}) \subset \widehat{C}$ (where $\widehat{C} = \{y \in \widehat{\mathbb{R}}^n | y_j \geq 0, \text{ for any } j\}$); moreover, this correspondence is one-to-one. Thus, corner singularities are identified with $\{J_1, \ldots, J_n\}$-equivariant singularities. Under this identification the form (1) corresponds to an invariant n-dimensional fold [8–10]:

$$\widehat{W}(y) = \sum \sigma_k y_k^4 + \sum_{q \geq 2} a_{k_1, \ldots, k_q} y_{k_1}^2 \cdots y_{k_q}^2 \tag{3}$$

and the form (2) to the evolute

$$\widehat{W}(y, \lambda) = \widehat{W}(y) + \sum \lambda_{k_1, \ldots, k_q} y_{k_1}^2 \cdots y_{k_q}^2, \tag{4}$$

which defines a miniversal deformation of the singularity (3) in the class of $\{J_1, \ldots, J_n\}$-invariant functions. Moreover, regular c.c. points of $W(\cdot, \lambda)$ correspond to the (usual) regular (critical) points of $\widehat{W}(\cdot, \lambda)$, and corresponding regular critical points have equal Morse indices (the Morse index of a regular c.c. point $a \in C$ is defined as the usual Morse index of the restriction $W|_{\mathbb{R}_K^n}$, $K = \{1, \ldots, n\} \setminus \operatorname{supp}(a)$, plus the number of negative first derivatives of $W(\cdot, \lambda)$ at a). The bifurcation diagrams of a corner singularity and of the corresponding $\{J_1, \ldots, J_n\}$-equivariant singularity are also identified.

Let Σ be the bifurcation diagram of functions (without the Maxwell stratum) for a corner singularity of the function (1) at the origin, i.e., Σ is the germ (at the origin) of the parameter value set in the base of a bounded miniversal deformation $W(x, \lambda)$ (obtained, for example, from (2) by the restriction $W(0, \lambda) = 0$), such that $W(\cdot, \lambda)$ has a degenerate c.c. point. For every $K \subset \{1, \ldots, n\}$ and every $j \notin K$, let $\Sigma_{K;j}$ denote the component of Σ consisting of the λ's corresponding to c.c. points x such that $\operatorname{supp}(x) \subset K$, $\operatorname{supp}(\operatorname{grad}_x W(x, \lambda)) \subset \{1, \ldots, n\} \setminus (K \cup j)$ For a semiregular singularity (1) (from now on we assume that $m = n$), the component $\Sigma_{K;j}$ is defined (for $K \neq \emptyset$) by the relations

$$\gamma_{K;j}(\lambda) = 0, \qquad \gamma_{K;k}(\lambda) \geq 0 \quad \forall k \in K, \tag{5}$$

where $\gamma_{K;k}$ is x_K expressed in terms of λ by means of the system of equations $\frac{\partial W}{\partial x_k} = x_r = 0$, $k \in K$, $r \notin K$, and $\gamma_{K;j}$ is obtained from $\frac{\partial W}{\partial x_j}(x, \lambda)$ by replacing x with its explicit expression in terms of λ from the same system of equations. Let γ_K denote the column vector with components $\gamma_{K;k}$ and H_K the matrix with elements $h_{p,q} = \frac{\partial^2 W(0,0)}{\partial x_p \partial x_q}$, $(p, q) \in K \times K$. Then

$$\gamma_K = -H_K^{-1} b_K + o(b_K), \tag{6}$$

where b_K is the column vector with components λ_K, $k \in K$ (H_K is invertible since 0 is a semiregular critical point of W). It is easy to see that

$$\gamma_{K;j}(\lambda) = \lambda_j + \sum_k h_{j,k}(\lambda) + o(b_K). \tag{7}$$

It follows from (5)–(7) that $\Sigma_{k;j}$ is a smooth submanifold of codimension 1 in the base of the deformation $W(x, \lambda)$.

THEOREM 1. *For any semiregular corner singularity* (1) (*with* $m = n$),

$$\Sigma = \bigcup_{K;j} \Sigma_{K;j}, \qquad K \subset \{1, \ldots, n\}, \ j \notin K. \tag{8}$$

THEOREM 2. *Formula* (6) *describes the asymptotic behavior* (*as* $\lambda \to 0$) *of a regular c.c. point with support K for a deformation* (2) (*of a semiregular singularity* (1)). *If $\beta_K(\lambda)$ is the corresponding critical value of the function* (2) *at this point, then*

$$\beta_K(\lambda) = -\frac{1}{2}\Sigma_{p,q}\overline{h}^K_{p,q}\lambda_p\lambda_q + o(|b_k|^2), \tag{9}$$

where $\overline{h}^K_{p,q}$ is the (p, q)-element of the inverse of H_K.

Soft bifurcations (i.e., bifurcations such that (Hx, x) is positive in C) are the most interesting ones for applications. The case $n = 2$ has been investigated fairly thoroughly [11–14]. Taking $\lambda_3 = h_{1,2} + \lambda_{1,2}$ in the deformation (3) (with $n = 2$), we obtain

$$W(x, \lambda) = x_1^2 + x_2^2 + 2\lambda_3 x_1 x_2 + \lambda_1 x_1 + \lambda_2 x_2, \qquad x \in \mathbb{R}^2, \ \lambda \in \mathbb{R}^3.$$

The diagram Σ is the union of the components $\Sigma_{\phi;1}, \Sigma_{\phi;2}, \Sigma_{2;1}, \Sigma_{1;2}$, where $\Sigma_{\phi;1}$ and $\Sigma_{\phi;2}$ are defined by the relations $\lambda_1 = 0$ and $\lambda_2 = 0$, and $\Sigma_{2;1}$ and $\Sigma_{1;2}$ by the relations $\lambda_1 - \lambda_3\lambda_2 = 0$, $\lambda_2 \leq 0$ and $\lambda_2 - \lambda_3\lambda_1 = 0$, $\lambda_1 \leq 0$. The intersection of all the components is a straight line, $\lambda_1 = \lambda_2 = 0$ (see Figure 1). The third and fourth components also intersect in a half-line, $\lambda_3 = 1$, $\lambda_1 = \lambda_2 \leq 0$, and both of them contain different "halves" of the line $\lambda_3 = -1$, $\lambda_1 + \lambda_2 = 0$. The first and third (second and fourth) components intersect in the ray $\lambda_1 = \lambda_3 = 0$, $\lambda_2 \leq 0$ (in the ray $\lambda_2 = \lambda_3 = 0$, $\lambda_1 \leq 0$). From the point of view of stratification of Σ over the stratum $\lambda_1 = \lambda_2 = 0$, singular points are those for which $\lambda_3 \in \{-1, 0, 1\}$. All degenerate critical points with $\lambda_1^2 + \lambda_2^2 \neq 0$ and $|\lambda_3| \neq 1$ have a boundary singularity of type B_2 [1]. The asymptotic formulas (7) are exact if $n = 2$. The coordinates of a c.c. point of second order (for first-order points everything is obvious) are given by

$$x_j = -\frac{\lambda_j - \lambda_3\lambda_k}{2(1 - \lambda_3^2)}, \qquad j \neq k.$$

The critical values can be obtained by using (9): $\beta_j(\lambda) = -\lambda_j^2/4$ (for first-order points), and

$$\beta_{1,2}(\lambda) = -\frac{\lambda_1^2 + \lambda_2^2 - 2\lambda_1\lambda_2\lambda_3}{4(1 - \lambda_3^2)}$$

(for second-order points). The topological metamorphoses of the level curves of the function $\widehat{W}(\cdot, \lambda)$ (where $\lambda \notin \Sigma$) are described in the following tables of contours (Figures 2–4).

Except for the case $\{\lambda_1 > 0, \lambda_2 > 0\}$, the level curves in the figures pass through saddle points. Minimum points are represented by dark points, max-

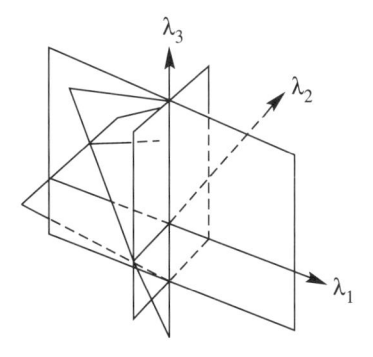

FIGURE 1

FIGURE 2

FIGURE 3

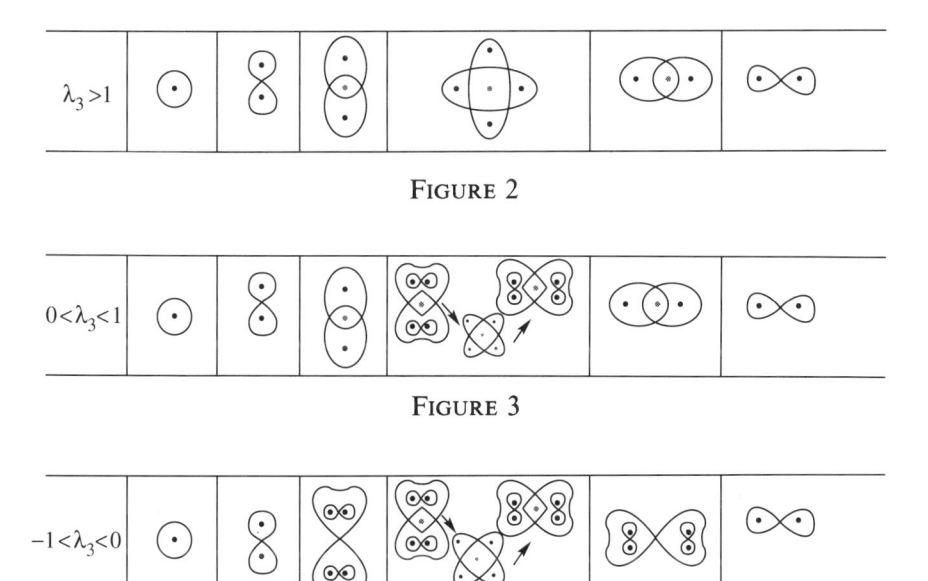

FIGURE 4

imum points by light ones. At present, the distribution of softly bifurcating points of conditional minima on the faces of a corner C has been described in the literature for $n = 3$.

§3. Symmetry breaking and adjacency of one-dimensional singularities

The multimodality of bifurcations of equilibrium states in physical systems is often due to the presence of a symmetry. If this symmetry is broken, phenomena may occur that are not characteristic of systems with symmetries. For example, cascade bifurcations may appear—branches of stationary states containing a sequence of births and deaths of stable states, accompanied by an overall drop in the values of the load parameter [15]. The number of steps of a cascade bifurcation is related to the multiplicity of the adjoining one-dimensional singularities. One-step bifurcations are generated by one-dimensional folds; two-dimensional folds with rectangular symmetry lead to three-step bifurcations. The latter are caused by the adjacency (of singularities) $X_9 \to A_7$ [16].

THEOREM 3. *The singularity adjoining the singularity of the function* (3) *at the origin is of type* A *and multiplicity* $k = 2^{n+1} - 1$.

The proof relies on the deformation $g(x_1^2, x_2^2 + \varepsilon x_1, \ldots, x_n^2 + \varepsilon x_{n-1})$, where g is a function expressing (3) in terms of the squares of variables [8].

Let M be the affine subvariety of the space of polynomials in $x_1, \ldots, x_n$ consisting of the polynomials

$$\sum \sigma_j x_j^4 + \sum_{k_j \leq 2, \sum k_j \geq 4} a_{k_1, \ldots, k_n} \xi_1^{k_1} \cdots \xi_n^{k_n}.$$

Let M_0 be the domain in M consisting of polynomials with quadratic parts of finite multiplicity (of multiplicity 3^n, [16]). These polynomials are called n-dimensional folds. The coordinates of the points of M are given by sets of coefficients $a = \{a_{k_1, \ldots, k_n}\}$. Following [17], we denote by $k(a)$ the maximal multiplicity of a singularity of type A adjoining the polynomial with coefficients $a = \{a_{k_1, \ldots, k_n}\}$.

THEOREM 4. *There exists an open dense subset of* M_0 *such that for any points in* M_0,

$$k(a) \leq \frac{n(n+1)}{2} + \frac{n(n+1)(n+2)}{6} < 2^{n+1} - 1.$$

For a proof along the lines presented in [17], see [8].

REMARK. If $n \geq 4$, we have

$$\frac{n(n+1)}{2} + \frac{n(n+1)(n+2)}{6} < 2^{n+1} - 1.$$

Hence the complex analogues of Theorems 3 and 4 and the results of [18] imply that the topological type of the bifurcation diagram along a stratum $\mu = $ const of a complex n-dimensional fold is not a constant.

§4. Examples

Various types of corner singularities arise in connection with the equilibrium of a longitudinally compressed rod with constraints against bending. If the ends of the rod are rigidly clamped, the equilibrium configurations are defined by the following boundary value problem [19]:

$$-(\ddot{x}(s) + \lambda \sin x(s)) + a(s, \varepsilon) = 0, \qquad x(0) = x(1) = 0,$$

where $x(s)$ is the angle between the vertical line and the tangent to the center line of the rod, s is a parameter measuring the length of the rod, $0 \leq s \leq 1$, λ is a parameter measuring the longitudinal compression force, $0 \leq \lambda \leq L$, $a(s, \varepsilon)$ is the function of initial imperfections, $a(s, 0) = 0$.

Thus the problem is simply the Euler-Lagrange equation for the extremals of the functional

$$V(x, \lambda, \varepsilon) = \frac{\|\dot{x}\|^2}{2} + \lambda \int_0^1 \cos x(s)\, ds + \langle a, x \rangle,$$

where $\|y\| = \sqrt{\langle y, y \rangle}$, $\langle a, x \rangle = \int_0^1 a(s, \varepsilon) x(s)\, ds$. If the constraints are represented by a system of inequalities $\langle \ell_k, x \rangle \geq 0$, $k \leq n$, with sufficiently smooth functions ℓ_k, then the key function

$$W(\xi, \lambda, \varepsilon) = \inf_{\langle \ell_j, x \rangle = \xi_j, \forall j} V(x, \lambda, \varepsilon)$$

is defined and smooth, and its quadratic part is defined by the quadratic part of V provided that $\|\dot{x}\|^2 > L\|x\|^2$ for every x, $x \perp \ell_j$, and all $j \leq n$.

As an example of a nonlinear boundary-value problem symmetric relative to a point of commuting involutions, we consider the homogeneous van Karman equation [20]

$$\Delta^2 \omega[\omega, \varphi] + \lambda \omega_{xx} = \Delta^2 \varphi + \frac{1}{2}[\omega, \omega] = 0$$

with boundary conditions $\varphi = \Delta \varphi = \omega = \Delta \omega = 0|_{\partial \Omega_a}$, corresponding to a plate attached to the boundary by a hinge. Here Δ is the Laplace operator, $[\omega, \varphi] = \omega_{xx}\varphi_{yy} + \omega_{yy}\varphi_{xx} - 2\omega_{xy}\varphi_{xy}$; ω is the deflection, φ the tension, $\Omega_a = [0, a] \times [0, 1]$, λ the load parameter. Eliminating φ from the equation and dilating the argument $x \to ax$, we obtain a boundary-value problem for the reduced von Karman equation:

$$a^4 \Delta_a^2 \omega + \frac{1}{2}[\omega, \Delta_a^{-2}[\omega, \omega]] + \lambda a^2 \omega_{xx} = 0,$$

$$\omega = \Delta \omega = 0|_{\partial \Omega}, \quad \Omega = [0, 1] \times [0, 1]$$

(where $\Delta_a = \frac{1}{a^2}\left(\frac{\partial}{\partial x}\right)^2 + \left(\frac{\partial}{\partial y}\right)^2$). The inversion Δ_a^{-1} is understood as the Green's operator $\psi \to \varphi$, where φ is a solution of the Poisson equation $\Delta_a \varphi = \psi$ [21]. This is the Euler-Lagrange equation for extremals of the functional

$$V(\omega, \lambda, a) = \frac{a^4}{2}\|\Delta_a \omega\|^2 - \frac{\lambda a^2}{2}\|\omega_x\|^2 + \frac{1}{8}\|\Delta_a^{-1}[\omega, \omega]\|^2$$

(e.g., in the space $E = C^{4+\alpha}(\Omega) \cap \{\omega = \Delta \omega = 0|_{\partial \Omega}\}$). Here $\|\omega\| = \sqrt{\langle \omega, \omega \rangle}$, $\langle \cdot, \cdot \rangle$ is the scalar product in $L_2(\Omega)$. The functional $V(\cdot, \lambda, a)$ is invariant under the involutions [20] $J_1: \omega(x, y) \to -\omega(1-x, y)$ and $J_2: \omega(x, y) \to \omega(1-x, y)$. For sufficiently small $\lambda - a\pi^2/2$, $a - \sqrt{2}$ and $(\omega, \xi) \in E \times \mathbb{R}^2$, the function $W(\xi, \lambda, a) = \inf_{\langle \ell_j, \omega \rangle = \xi_j, \forall j} V(\omega, \lambda, a)$, $\xi \in \mathbb{R}^2$, is defined and smooth [22], where $\ell_k = 2 \sin \pi k x \sin \pi y$, $k = 1, 2$. If follows from calculations published in [23, 24] that in this case $h_{1,2} > \sqrt{h_{1,1} \cdot h_{2,2}}$. Hence the topology of the level curves of the key function $W(\cdot, \lambda, a)$ is that shown in Figure 2.

References

1. V. I. Arnol'd, *Critical points of functions on a manifold with boundary, the simple Lie groups B_k, C_k, F_4 and singularities of evolutes*, Uspekhi Mat. Nauk **33** (1978), no. 5, 91–105; English transl. in Russian Math. Surveys **33** (1978), no. 5.
2. D. Siersma, *Singularities of functions on boundaries, corners, etc.*, Quart. J. Math. **32** (1981), 119–127.
3. C. T. C. Wall, *A note on symmetry of singularities*, Bull. London Math. Soc. **12** (1980), no. 3, 169–175.
4. V. Poénaru, *Singularités C^∞ en présence de symétrie. En particulier en présence de la symétrie d'un groupe de Lie compact*, Lecture Notes in Math., vol. 510, Springer-Verlag, Berlin, Heidelberg, and New York, 1976, pp. 61–85.
5. V. A. Trenogin and N. A. Sidorov, *Investigation of bifurcation points and nontrivial solutions of nonlinear equations*, Differential and Integral Equations, vol. 1, Izdat. Irkutsk. Univ., Irkutsk, 1972, pp. 216–247. (Russian)
6. M. A. Krasnosel'skiĭ, N. A. Bobylëv, and È. M. Mukhamadiev, *On a scheme for the investigation of degenerate extremals of functionals of classical variational calculus*, Dokl. Akad Nauk SSSR **240** (1978), no. 3, 530–533; English transl. in Soviet Math. Dokl.
7. T. E. Marsden, *On the geometry of the Liapunov-Schmidt procedure*, Lecture Notes in Math., vol. 755, Springer-Verlag, Berlin and New York, 1979, pp. 77–82.
8. Yu. I. Sapronov, *Spherical symmetry breaking in nonlinear variational problems*, Analysis on Manifolds and Differential Equations, Izdat. Voronezh. Gos. Univ., Voronezh, 1986, pp. 88–111. (Russian)
9. _____, *On the topology of the bifurcation diagram of a multidimensional fold*, Baku International Topology Conference, Abstracts of Lectures, 1987, p. 271. (Russian)
10. _____, *Corner singularities and multidimensional folds*, Funktsional. Anal. i Prilozhen. **22** (1988), no. 3, 85–86; English transl. in Functional Anal. Appl. **22** (1988), no. 3.
11. T. Poston and I. Stewart, *Catastrophe theory and its applications*, Pitman, London, 1978.
12. R. Gilmore, *Catastrophe theory for scientists and engineers*, Wiley, New York, 1981, 285 pp.
13. A. D. Bryuno, *A study of the limited three-body problem. I, Periodic solutions of Hamilton systems*, Akad. Nauk SSSR Inst. Prikl. Mat. Preprint, 1978, 44 pp. (Russian)
14. V. I. Arnol'd, V. V. Kozlov, and A. I. Neustadt, *Mathematical aspects of classical and celestial mechanics*, Itogi Nauki i Tekhniki. Sovremennye Problemy Matematiki. Fundamental'nye Napravleniya, vol. 3, VINITI, Moscow, 1985, pp. 5–304; English transl. in Encyclopedia Math. Sci., vol. 1, Springer-Verlag, Berlin, 1989.
15. Yu. I. Sapronov, *Multimodal bifurcations of elastic equilibria*, Prikl. Mat. Mekh. **52** (1988), no. 6, 997–1006; English transl. in J. Appl. Math. Mech. **6** (1972).
16. V. I. Arnol'd, *Normal forms of functions in neighborhoods of degenerate critical points*, Uspekhi Mat. Nauk **29** (1974), no. 2, 11–49; English transl. in Russian Math. Surveys **29** (1974).
17. S. M. Guseĭn-Zade and N. N. Nekhoroshev, *On adjacency of singularities to points of stratum $\mu = const$ of a singularity*, Funktsional. Anal. i Prilozhen. **17** (1983), no. 4, 82–83; English transl. in Functional Anal. Appl. **17** (1983), no. 4.
18. A. N. Varchenko and S. M. Guseĭn-Zade, *Topology of caustics, wave fronts and degenerate critical points*, Uspekhi Mat. Nauk **39** (1984), no. 2, 190–191; English transl. in Russian Math. Surveys **39** (1984).
19. A. Love, *A treatise of the mathematical theory of elasticity*, 4th ed., Cambridge Univ. Press, London, 1927; reprint, Dover, New York, 1944.
20. A. L. Vanderbauwhede, *Generic bifurcation and symmetry with an application to the von Karman equations*, Proc. Royal Soc. Edinburgh **81A** (1978), 211–235.
21. E. A. Volkov, *On solutions of boundary problems for the Poisson equation in a rectangle*, Dokl. Akad Nauk SSSR **147** (1962), no. 2, 13–16; English transl. in Soviet Math. Dokl. **3** (1962).

22. Yu. I. Sapronov, *On bimodal bifurcations of solutions of the von Karman equation*, Abstracts of Fourth Republican Conference "Nonlinear Problems of Math. Physics", Donetsk, 1987, p. 132.
23. G. H. Knightly, *Some mathematical problems from plate and shell theory*, Lecture Notes in Pure and Appl. Math., vol. 19, Dekker, New York, 1977, pp. 245–268.

Translated by N. ZOBIN

Newton Polyhedra (Algebra and Geometry)

A. G. KHOVANSKIĬ

Newton polyhedra establish a relationship between algebraic geometry and the geometry of polyhedra. In this paper we discuss this relationship and its applications in algebra and geometry.

§1. Computation of discrete invariants in terms of Newton polyhedra

In this section we discuss the computation of discrete algebraic invariants in geometric terms and, in particular, the computation of the number of roots of a system of equations in terms of the mixed volumes of polyhedra. An outline of an algebraic proof of the isoperimetric inequality will be given.

1.1. The ideology of general position. Suppose that the outcome of some natural process, such as a physical experiment, is the graph of a function of one variable over a closed interval of finite length. One feels that such a function ought to have only finitely many roots. Can this be proved rigorously? "Classical mathematics" answers this question with an univocal "no": for any given closed subset of a closed interval (such as a Cantor perfect set) there exists an infinitely differentiable function whose set of zeroes is the given set. In the "mathematics of general position", however, the statement can be proved: The space of smooth functions over a closed interval contains a dense open set of functions which have finitely many zeroes, maxima, and minima. A "generic" function is a function from a dense open set. "Nongeneric" functions form a hypersurface of codimension one in the space of functions. Various pathological effects, such as the existence of infinitely many zeroes over a closed interval of finite length, may of course occur, but for the overwhelming majority of functions no complications arise. It is natural to expect that a function occurring in some natural context (and not constructed specially as a counterexample) will be a function in general position.

For various classes of objects (functions, mappings, differential equations) one can often single out those in general position, which, first, constitute the overwhelming majority, and, second, behave in a much simpler way than an arbitrary object.

The main task of singularity theory is to identify and investigate objects in general position. To illustrate how fruitful this approach can be, let us compare the following two assertions:

1) There is no algebraic algorithm that enables us to determine, using only the coefficients of a differential equation, whether or not its equilibrium position is stable.

2) For "generic" differential equations there is an explicit (algebraic) procedure that will tell us whether an equilibrium is stable or not.

Thus, the unsolvability of the stability problem, as declared by "classical mathematics", pertains only to "special" equations, which should never be encountered in real life except, of course, under special circumstances. Had mathematicians unquestioningly accepted the quite convincing theorem that there is no algorithm for the stability problem and failed to investigate the case of general position, we would have remained ignorant of a remarkable criterion for stability which has innumerable applications and may be found in any textbook on differential equations.

1.2. **Complex and real cases in the "mathematics of general position".** In both complex and real cases, exceptional objects usually form a hypersurface. In the complex case, however, the hypersurface is also complex and its real codimension is two (one complex equation can be considered as two real ones—a complex number is zero if both its real and imaginary parts are zero). A subset of codimension two does not divide the space into parts, and that is why, starting from one generic complex object, one can reach any other without passing through degenerate, nongeneric objects. Discrete invariants remain unchanged in the process—hence the discrete invariants of any two complex generic objects are equal. Not so in the real case: a generic real hypersurface divides the space into parts and even discrete invariants of different generic real objects may be distinct. Here is a simple example: in the space of polynomials of degree n in one complex variable, the nongeneric polynomials having multiple roots form a complex hypersurface. The number of complex roots of a generic complex polynomial of degree n does not depend on the choice of the polynomial—it is n. On the other hand, the number of real roots of a generic real polynomial depends on the choice of the polynomial (it may be n, $n-2$, $n-4$, ...).

1.3. **Examples of computation with Newton polyhedra.** Newton polyhedra generalize the notion of the degree of a polynomial and play a similar role. The Newton polyhedron of a polynomial of n variables is a polyhedron in the real linear space $\mathbb{R}^n$, defined as the convex hull of the points whose

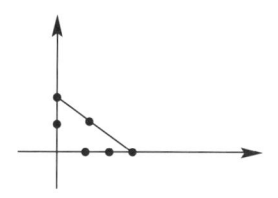

FIGURE 1

coordinates are the exponents of the powers that occur in the polynomial with nonzero coefficients.

EXAMPLES. 1. Consider the following polynomial p in x and y:

$$y^2 + Q_3(x) = y^2 + a_0 + a_1 x + a_2 x^2 + a_3 x^3.$$

The Newton polygon $\Delta(p)$ of this polynomial is shown in Figure 1. The point $(0, 2)$ in this figure corresponds to the monomial y^2, the points $(0, 0)$, $(0, 1)$, $(0, 2)$, and $(0, 3)$ to the monomials a_0, $a_1 x$, $a_2 x^2$, and $a_3 x^3$, respectively.

2. The Newton polyhedron of a generic polynomial of degree m in two variables is the simplex with vertices $(m, 0)$, $(0, m)$, $(0, 0)$.

The Newton polyhedron of a generic polynomial of degree m in n variables is the n-dimensional simplex $x_1 \geq 0, \ldots, x_n \geq 0$, $\sum x_i \leq m$. Thus we see that Newton polyhedra indeed generalize the notion of degree.

Discrete parameters of the complex variety defined by a generic equation $P(x) = 0$ with fixed Newton polyhedron $\Delta(P)$ do not depend on the specific choice of the equation; they depend only on $\Delta(P)$.

EXAMPLE. Consider a generic complex curve $P(x, y) = 0$ in the plane, with fixed Newton polyhedron Δ. Its genus is equal to the number of points with integer coordinates in the interior of the Newton polyhedron Δ [1]. Thus, the curve $y^2 + Q_3(x) = 0$, where $Q_3(x)$ is a sufficiently generic polynomial of third degree, has genus 1, since its Newton polygon contains only one integral point (with coordinates $(1, 1)$; see Figure 1). Similarly, the genus g of a generic curve $P_m(x, y)$ of degree m, is $g = (m - 1)(m - 2)/2$, which is just the number of points in the interior of the simplex $x_1 \geq 0$, $x_2 \geq 0$, $x_1 + x_2 \leq m$.

The formula for the genus of a generic curve is well known. We see that computations with Newton polyhedra generalize the classical computations of invariants of algebraic varieties, which are defined by generic equations of fixed degrees, and give answers a geometrical meaning. Here is a multidimensional generalization of the previous example.

EXAMPLE. The arithmetical genus of the hypersurface defined in $\mathbb{C}^n$ by a generic equation $P(x) = 0$ with fixed Newton polyhedron is equal to the number of points with integer coordinates in the interior of the Newton polyhedron [1]. The corresponding classical computation: the arithmetical genus of a general hypersurface of degree m in n-dimensional space is

$(m-1)\cdots(m-n)/n!$; this number is just the number of points with integer coordinates inside the simplex $x_1 \geq 0, \ldots, x_n \geq 0$, $\sum x_i \leq m$.

As a rule, the answers become simpler and more symmetric if one considers solutions of equations not in $\mathbb{C}^n$, but in $(\mathbb{C} \setminus 0)^n$, that is, the subdomain of $\mathbb{C}^n$ obtained by deleting all the coordinate hyperplanes.

EXAMPLE [1]. The Euler characteristic of the hypersurface defined in $(\mathbb{C} \setminus 0)^n$ by a generic equation $P(x) = 0$ with fixed Newton polyhedron is equal to the volume of the polyhedron multiplied by $(-1)^{n-1}n!$. (The Euler characteristic of a hypersurface in $\mathbb{C}^n$ is given by a more complicated formula. However, since Euler characteristic is additive, this more complicated formula reduces to the previous formula.)

As simple an object as the volume of the convex hull of a system of points is an extremely complicated function of the coordinates. Such objects, considered in the usual numerical terms, are so cumbersome that no progress would be possible without a knowledge of their geometrical interpretation.

1.4. Number of roots of n equations in n unknowns. By the Bézout Theorem, the number of nonzero roots of a polynomial f in one variable is equal to the difference between the maximum and minimum degrees of the monomials occurring in the polynomial. The Newton polyhedron of the polynomial f is a closed interval and the above-mentioned difference is the length of this interval. This observation is generalized to the multidimensional case in the following theorems.

KUSHNIRENKO'S THEOREM [2]. *The number of solutions in $(\mathbb{C} \setminus 0)^n$ of a generic polynomial system of n equations $P_1 = \cdots = P_n = 0$ with identical Newton polyhedra $\Delta(P_1) = \cdots = \Delta(P_n) = \Delta$ is equal to the volume $V(\Delta)$ of the Newton polyhedron multiplied by $n!$.*

EXAMPLE. As already stated, the Newton polyhedron of a polynomial of degree m in n variables is a simplex $x_1 \geq 0, \ldots, x_n \geq 0$, $\sum x_i \leq m$ (on the assumption that the polynomial contains all monomials of degrees $\leq m$). The volume of this simplex is $m^n/n!$. By Kushnirenko's Theorem, the number of roots of a general system of n equations of degree m in n unknowns is m^n.

This answer agrees with the Bézout Theorem. The Newton polyhedron of a polynomial that does not contain all monomials of degree $\leq m$ may be smaller than a simplex, and the number of solutions according to Kushnirenko's Theorem may be less than the number m^n given by the Bézout Theorem. The reason is that, as some monomials do not appear, some points at infinity may turn out to be roots. Since the Bézout Theorem computes the number of roots in the projective space, it counts these "parasitic" roots, while Kushnirenko's Theorem does not.

BERNSTEIN'S THEOREM [3, 1]. *The number of solutions in $(\mathbb{C} \setminus 0)^n$ of a generic system of n polynomial equations $P_1 = \cdots = P_n = 0$ with Newton* polyhedra $\Delta_1, \ldots, \Delta_n$ is equal to the mixed volume $V(\Delta_1, \ldots, \Delta_n)$ multiplied by $n!$.

We first define mixed volume.

By the *Minkowski sum* of two subsets of a linear space we mean the set of all sums of pairs of vectors, one from the first subset and another from the second one. The product of a subset by a number is defined similarly. The Minkowski sum of convex bodies (convex polyhedra, convex polyhedra with vertices at points with integer coordinates) is a convex body (convex polyhedron, convex polyhedron with vertices at points with integer coordinates). The role of Minkowski summation in the theory of Newton polyhedra is due to the following simple proposition: the Newton polyhedron of a product of polynomials is the Minkowski sum of their Newton polyhedra.

MINKOWSKI'S THEOREM. *The volume of a body that is a linear combination with positive coefficients of fixed convex bodies in $\mathbb{R}^n$ is a homogeneous polynomial of degree n in the coefficients of the linear combination.*

DEFINITION. The *mixed volume* $V(\Delta_1, \ldots, \Delta_n)$ of convex bodies $\Delta_1, \ldots, \Delta_n$ in $\mathbb{R}^n$ is the coefficient of $\lambda_1 \cdots \lambda_n$ in the polynomial $V(\lambda_1 \Delta_1 + \cdots + \lambda_n \Delta_n)$, divided by $n!$, where $V(\Delta)$ is the volume of Δ.

The mixed volume of n identical bodies is equal to the volume of any one of them. The mixed volume of n bodies is expressed in terms of usual volumes of their sums in the same way as the product of n numbers is expressed in terms of the nth powers of their sums. For example, for $n = 2$ we have

$$ab = \frac{1}{2}\left[(a+b)^2 - a^2 - b^2\right],$$

$$V(\Delta_1, \Delta_2) = \frac{1}{2}\left[V(\Delta_1 + \Delta_2) - V(\Delta_1) - V(\Delta_2)\right].$$

Similarly, for $n = 3$,

$$V(\Delta_1, \Delta_2, \Delta_3) = \frac{1}{3!}\left[V(\Delta_1 + \Delta_2 + \Delta_3) - \sum_{i<j} V(\Delta_i + \Delta_j) + \sum_i V(\Delta_i)\right].$$

EXAMPLE. Let Δ_1 be the rectangle $0 \leq x \leq a$, $0 \leq y \leq b$, and Δ_2 the rectangle $0 \leq x \leq c$, $0 \leq y \leq d$. The Minkowski sum $\Delta_1 + \Delta_2$ is the rectangle $0 \leq x \leq a+c$, $0 \leq y \leq b+d$. The mixed volume $V(\Delta_1, \Delta_2)$ is $ad + bc$. The number $ad + bc$ is the permanent of the matrix $\begin{pmatrix} a & b \\ c & d \end{pmatrix}$ (the definition of the permanent is the same as that of the determinant, except that all summands enter with the plus sign). In the multidimensional case the mixed volume of n parallelepipeds whose sides are parallel to the coordinate axes is also equal to the permanent of the corresponding matrix.

EXAMPLE. The number of roots of a generic system of polynomial equations in which the ith variable occurs in the jth equation to power at most $a_{i,j}$ is equal to the permanent of the matrix $(a_{i,j})$ times $n!$.

For equations with identical Newton polyhedra, the statements of Kushnirenko's and Bernstein's theorems coincide.

Newton polyhedra can be used to compute not only discrete invariants of hypersurfaces and numbers of solutions of systems of n equations in n unknowns, but also discrete invariants of the varieties of solutions of systems of k equations in n unknowns. Here is a very simple example.

EXAMPLE [4]. For $0 \le i < n-k$, the i-dimensional homology group of the variety X of solutions of a generic system of k polynomial equations $P_1 = \cdots = P_k = 0$ in $(\mathbb{C}\setminus 0)^n$ is isomorphic to the i-dimensional homology group of the space $(\mathbb{C}\setminus 0)^n$, if the Newton polyhedra $\Delta_1, \ldots, \Delta_k$ of the system have full dimension, i.e., $\dim\Delta_1 = \cdots = \dim\Delta_k = n$. Hence, $\dim H_i(X) = \binom{n}{i}$, for $0 \le i < n-k$. In particular, if $k < n$, the variety X is connected.

Let us return now to mixed volumes.

1.5. Aleksandrov-Fenchel inequalities. Many geometrical invariants can be expressed in terms of mixed volumes of bodies. For example, the ε-neighborhood of a convex body Δ is the set $\Delta + \varepsilon K$, where K is the unit ball about the origin. By the Minkowski Theorem, the volume of the ε-neighborhood of a body Δ is the following polynomial in ε:

$$V(\Delta) + n\varepsilon V_{n-1}(\Delta) + \frac{n(n-1)}{2}\varepsilon^2 V_{n-2}(\Delta) + \cdots,$$

where

$$V_l(\Delta) = V(\underbrace{\Delta, \ldots, \Delta}_{l \text{ times}}, \underbrace{K, \ldots, K}_{n-l \text{ times}}).$$

On the other hand, it is clear that in the first approximation with respect to Δ the volume of the ε-neighborhood of Δ is $V(\Delta) + \varepsilon S(\Delta)$, where $S(\Delta)$ is the surface area of Δ. From this we conclude that the surface area of Δ is $nV_{n-1}(\Delta)$. (It is not difficult to show that the mixed volume $V_l(\Delta)$ coincides, up to a factor depending only on l and n with the mean value of the area of the projection of Δ onto an l-dimensional plane.)

The following remarkable inequality of Aleksandrov-Fenchel for mixed volumes of bodies is a generalization of the isoperimetric inequality

$$V^2(K_1, K_2, \ldots, K_n) \ge V(K_1, K_1, \ldots, K_n) \cdot V(K_2, K_2, \ldots, K_n)$$

where $K_1, \ldots, K_n$ are nonempty convex bodies. The following assertion is a simple formal corollary of these inequalities: for any two convex bodies K_1 and K_2 and any integer l, $0 < l < n$,

$$V^n(\underbrace{K_1, \ldots, K_1}_{l \text{ times}}, \underbrace{K_2, \ldots, K_2}_{n-l \text{ times}}) \ge V^l(K_1) \cdot V^{n-l}(K_2).$$

Applying this assertion to the case when K_1 is Δ, K_2 is the unit ball K, and $l = n - 1$, we obtain the isoperimetric inequality:

$$S(\Delta) \ge \frac{1}{n} V^{(n-1)/n}(\Delta) \cdot V^{1/n}(K).$$

The isoperimetric inequality furnishes an upper bound for the volume of a body in terms of its surface area. This bound is best possible: for a ball the equality sign holds.

The proof of the Aleksandrov-Fenchel inequalities is far from simple. Three geometrical proofs of these inequalities are known. Two were published by A. D. Aleksandrov in 1936 [5]. One of them is combinatorical (the proof is worked out for polyhedra and then extended to all convex bodies by continuity). The other is analytical; it uses the theory of selfadjoint elliptic operators depending on parameters (the proof is worked out for smooth, strictly convex bodies and then extended to all convex bodies by continuity). Fenchel's geometrical proof also dates to 1936 (it is so complicated that it has never been published in full).

1.6. An algebraic proof of the Aleksandrov-Fenchel inequalities. Thanks to the link created by Newton polyhedra between algebraic geometry and the geometry of polyhedra, algebraic theorems of general type yield considerable information about the geometry of polyhedra. As an example, we will show how the Hodge Index Theorem can be used to obtain a simple proof of the Aleksandrov-Fenchel Theorem. The proof outlined below was found independently in 1979 by Teissier [6, 7] and the author [8, 9].

We first recall the Hodge Inequality from the algebraic geometry of surfaces [10], which will play a key role in what follows. Let Γ_1 and Γ_2 be two complex curves on a compact connected complex algebraic surface F. Let the self-intersection number of one of the curves be positive. The Hodge Inequality states that

$$\langle \Gamma_1, \Gamma_2 \rangle^2 \geq \langle \Gamma_1, \Gamma_1 \rangle \langle \Gamma_2, \Gamma_2 \rangle,$$

where $\langle \Gamma_1, \Gamma_2 \rangle$ is the intersection number of the curves Γ_1 and Γ_2, $\langle \Gamma_1, \Gamma_1 \rangle$ and $\langle \Gamma_2, \Gamma_2 \rangle$ are self-intersection numbers of these curves. (Recall that the self-intersection number of a curve is defined as the intersection number of the curve with its image under a slight deformation.)

The Aleksandrov-Fenchel inequalities will be inferred from the Hodge Inequality and Bernstein's Theorem, as follows.

Consider the noncompact algebraic surface defined in $\mathbb{C}^n$ by a generic system of $n-2$ polynomial equations $P_3 = 0, \ldots, P_n = 0$ with Newton polyhedra $\Delta_3, \ldots, \Delta_n$ of full dimension $\dim \Delta_3 = \cdots = \dim \Delta_n = n$. Since by assumption the Newton polyhedra are of full dimension, the surface will be connected. Let A_1 and A_2 be two curves on this surface, defined by generic polynomial equations $P_1 = 0$ and $P_2 = 0$ with Newton polyhedra Δ_1 and Δ_2. The number of points of intersection of the curves A_1 and A_2 on our noncompact surface is equal to the number of solutions of the system of equations $P_1 = P_2 = P_3 = \cdots = P_n = 0$. By Bernstein's Theorem, this number is equal to $n! V(\Delta_1, \Delta_2, \Delta_3, \ldots, \Delta_n)$. Together with the curve A_1, let us consider a slightly deformed curve A_1', defined by an equation $P_1' = 0$ that contains the same monomials as the equation $P_1 = 0$ but with slightly different coefficients (in particular, the polynomials P_1 and P_1' have identical Newton polyhedra). Again by Bernstein's Theorem, the number of points of intersection of the curves A_1 and A_1' is $n!V(\Delta_1, \Delta_1, \Delta_3, \ldots, \Delta_n)$. Similarly, we construct a curve A_2' that intersects the curve A_2 at $n!V(\Delta_2, \Delta_2, \Delta_3, \ldots, \Delta_n)$ points.

The next step is compactification of the noncompact surface $P_3 = \cdots = P_n = 0$.

There is a special compactification F of this surface under which the closures Γ_1, Γ_1', Γ_2, and Γ_2' of the noncompact curves A_1, A_1', A_2, and A_2' do not intersect "at infinity". Such compactifications play an important technical role in the theory of Newton polyhedra. Some idea of such compactifications will be given in the next section. Since the curves $\Gamma_1 = \overline{A}_1$ and $\Gamma_1' = \overline{A}_1'$ have no points of intersection "at infinity", it follows that $\Gamma_1 \cap \Gamma_1' = A_1 \cap A_1'$; similar equalities hold for the other pairs of curves. Hence the intersection and self-intersection numbers of the curves Γ_1 and Γ_2 are determined by the formulas

$$\langle \Gamma_1, \Gamma_2 \rangle = n!V(\Delta_1, \Delta_2, \Delta_3, \ldots, \Delta_n),$$
$$\langle \Gamma_1, \Gamma_1 \rangle = n!V(\Delta_1, \Delta_1, \Delta_3, \ldots, \Delta_n),$$
$$\langle \Gamma_2, \Gamma_2 \rangle = n!V(\Delta_2, \Delta_2, \Delta_3, \ldots, \Delta_n).$$

Substituting these formulas into Hodge's Inequality, we obtain the Aleksandrov-Fenchel inequality for the polyhedra $\Delta_1, \Delta_2, \Delta_3, \ldots, \Delta_n$:

$$V^2(\Delta_1, \Delta_2, \Delta_3, \ldots, \Delta_n) \geq V(\Delta_1, \Delta_1, \Delta_3, \ldots, \Delta_n)V(\Delta_2, \Delta_2, \Delta_3, \ldots, \Delta_n).$$

It follows that the Aleksandrov-Fenchel inequality holds for polyhedra of full dimension with vertices at rational points (by a scale change one can place the vertices at points with integer coefficients). To complete the proof, it remains to use the continuity of mixed volume.

1.7. Minding's method. We conclude this section with a brief discussion of Bernstein's Theorem. To be precise, we will explain to the reader why it is possible to compute the number of roots of a system of two polynomial equations $f(x, y) = g(x, y) = 0$ using the Newton polygons of polynomials f and g. The method that will be discussed below is due to Ferdinand Gottlieb Minding (1806–1865). Minding expressed the answer in terms of the geometrical characteristics of polygons [11]. However, he was not familiar with the notion of mixed volume—when his paper was published (1841), Minkowski had not yet been born; he was therefore unable to extend his theorem to the multi-dimensional case, and the generalization was accomplished only in 1975 by D. N. Bernstein. We now outline Minding's method. Minding eliminates the variable y from the system $f(x, y) = g(x, y) = 0$. To that end, he considers the multi-valued function $y(x)$ defined by the equation $f(x, y) = 0$ and substitutes it into the second equation. The number of branches of the algebraic function $y(x)$ is equal to the degree of the polynomial f with respect to the variable y. Suppose that the degree is k, and let $y_1(x), \ldots, y_k(x)$ be the different branches of the function $y(x)$. The product of all k branches $g(x, y_i(x))$ of the multi-valued function $g(x, y(x))$ gives a function $p(x) = \prod_{1 \leq i \leq k} g(x, y_i(x))$; this function is, however, single-valued. It is clear that the zeroes of the function p correspond to the roots of the original system. Since p is a single-valued algebraic function, it is rational. Moreover, if the curve $f(x, y) = 0$ has no vertical asymptotes, then $p(x)$ will never become infinite for finite x, so it is a polynomial. Therefore, to determine the number of roots of the system we must find the number of roots of the polynomial $p(x)$, which is simply its degree. To determine the degree of the polynomial p, Minding suggests the following method. First compute the leading terms of the expansions of the branches $y_i(x)$ of $y(x)$ in fractional powers of x (the so-called Puiseux series) as $x \to \infty$. Then, substituting the leading terms of the expansions of $y_i(x)$ into $g(x, y)$, compute the leading terms of the expansions of the branches $g(x, y_i(x))$ and determine their (fractional) degrees. The required degree of the polynomial p is equal to the sum of the degrees of the branches $g(x, y_i(x))$ (this sum is always an integer).

Now Newton invented "his" polygons for the very purpose of determining the leading terms of expansions of algebraic functions in Puiseux series. Minding uses the Newton polygon of the polynomial $f(x, y)$ to determine the leading terms in the expansions of branches $y_i(x)$. He points out that in general position, when the coefficients of polynomials f and g are not bound by special relationships, the leading terms of the branches $y_i(x)$ do not cancel out after being substituted into $g(x, y)$. In addition, the degree of any branch $g(x, y_i(x))$ depends only on that of $y_i(x)$ and on the Newton polygon of the polynomial $g(x, y)$. It is noteworthy that Minding's algorithm for computing the number of solutions of the system $f(x, y) = g(x, y) = 0$ as the degree of the polynomial $p = \prod g(x, y_i(x))$, obtained by expanding the branches $y_i(x)$ in Puiseux series, also works in the degenerate case. In the case of degeneration, however, leading terms may cancel out, so that the number of roots and the degree of the polynomial p can decrease.

§2. Toric varieties and the combinatorics of polyhedra

In this section we shall give an intuitive idea of toric varieties and describe an application of the algebraic geometry of toric varieties to the combinatorics of polyhedra.

2.1. **The notion of toric variety.** It is well known that generic curves of degree m are far more easily and conveniently studied not in the plane $\mathbb{C}^2$, but in the projective plane $\mathbb{CP}^2$. This is because, first, the projective plane is

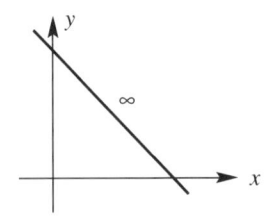

FIGURE 2

compact and, second, the generic curve of degree m in the projective plane is smooth (compactification introduces no singularities "at infinity").

Figure 2 is a schematic representation of the projective plane, showing the x and y axes and the line at infinity.

Now this picture clearly looks like the Newton polygon of a generic polynomial of degree m. Moreover, the monomials of the highest degree m on the side $x_1 + x_2 = m$, $x_1 \geq 0$, $x_2 \geq 0$ of the Newton polygon play the main role when one approaches the line at infinity. The monomials on the side $x_1 = 0$, $x_2 \geq 0$, $x_1 + x_2 \leq m$ of the Newton polygon dominate near the x axis, while monomials on the side $x_1 \geq 0$, $x_2 = 0$, $x_1 + x_2 \leq m$ are dominant near the y axis.

In the same way, a curve defined by a generic equation with a fixed Newton polygon Δ is more easily and conveniently considered in a special compactification of the space $(\mathbb{C} \setminus 0)^2$ based on the polygon Δ. To each side of the polygon Δ corresponds a deleted straight line $\mathbb{C} \setminus 0$, which is added to $(\mathbb{C} \setminus 0)^2$ during the compactification, and to each vertex of the polygon Δ corresponds an added point. The monomials on any one side of the Newton polygon become dominant when one approaches the corresponding deleted line, and the monomials at the vertices of the Newton polygon dominate when one approaches the corresponding points.

In the multi-dimensional case we can also construct compactifications of the space $(\mathbb{C} \setminus 0)^n$ based on Newton polyhedra [12]. They play the same role as the projective compactification for general varieties of fixed degrees. All these compactifications are what is known as toric varieties [13].

The space $(\mathbb{C} \setminus 0)^n$ is an algebraic group with respect to coordinatewise multiplication of vectors, known as an n-dimensional torus. A connected n-dimensional algebraic variety (in general, singular) on which an n-dimensional torus acts algebraically and has one orbit isomorphic to an n-dimensional torus, is called a *toric variety*. Under the action of a torus, a toric variety is broken up into a finite number of orbits isomorphic to tori of different dimensions. To every Newton polyhedron Δ we can associate a compact projective toric variety in such a way that every k-dimensional face of Δ corresponds to a complex k-dimensional orbit in the toric variety. If one face of the polyhedron is contained in the closure of another face, then the orbit corresponding to the first face is contained in the closure of the orbit corresponding to the second one. We will not present the construction of the toric variety associated with a Newton polyhedron Δ, but restrict ourselves to the following example.

EXAMPLE. The torus $(\mathbb{C} \setminus 0)^2$ acts on the projective plane $\mathbb{CP}^2$ by the following formula: an element $(u_1, u_2) \in (\mathbb{C} \setminus 0)^2$ takes the point $z_1 : z_2 : z_3$ into the point $u_1 z_1 : u_2 z_2 : z_3$. With this action, $\mathbb{CP}^2$ becomes a two-dimensional toric variety. This variety corresponds to the Newton polygon of a generic polynomial of any fixed degree $m > 0$. The sides of the polygon along the lines $x_1 = 0$, $x_2 = 0$, and $x_1 + x_2 = m$ correspond to the one-dimensional orbits $z_1 = 0$, $z_2 \cdot z_3 \neq 0$; $z_2 = 0$, $z_1 \cdot z_3 \neq 0$; and $z_3 = 0$, $z_1 \cdot z_2 \neq 0$. The vertices $(0, 0)$, $(m, 0)$, and $(0, m)$ correspond to the zero-dimensional orbits $(0:0:1)$, $(1:0:0)$, and $(0:1:0)$.

2.2. Simple polyhedra. We now turn to the combinatorics of polyhedra. A bounded polyhedron is said to be *simple* if it is an intersection of half-spaces in general position. An n-dimensional simple polyhedron has the same structure near each vertex as the positive orthant in $\mathbb{R}^n$ near the origin. In particular, each vertex of a simple n-dimensional polyhedron is incident with exactly n edges, and any k of these edges belong to one k-dimensional face containing the vertex.

REMARK. In "classical mathematics" a polyhedron can be defined either as the intersection of a finite number of subspaces or as the convex hull of a finite set of points. A priori, these two definitions determine dual objects, and in the mathematics of "general position" they are indeed distinct. As envisaged by the first definition, a generic polyhedron is a simple polyhedron. From the point of view of the second definition, it is a polyhedron all of whose proper faces are simplices. However, the theory of simple polyhedra differs only slightly from the theory of polyhedra with simplicial faces: polyhedra that are dual to simple polyhedra have simplicial faces, and vice versa.

The F-polynomial of a simple n-dimensional polyhedron is defined as the generating polynomial of the sequence formed by the numbers of faces of different dimensions in the polyhedron, i.e., $F(t) = \sum F_k t^k$, where F_k is the number of k-dimensional faces of the polyhedron.

By no means all polynomials are F-polynomials of simple polyhedra. Conditions for a polynomial to have this property were found by McMullen [14]. The necessity of these conditions was proved by Stanley [15], their sufficiency by Billera and Lee [16]. To formulate the conditions we need the notion of the H-polynomial of a polyhedron. To each F-polynomial we associate an H-polynomial defined by the formula $H(t) = F(t-1)$; i.e., if $F(t) = \sum F_k t^k$, then $H(t) = \sum H_k t^k$, where $H_k = \sum (-1)^{m-k} \binom{m}{k} F_k$.

We now present McMullen's conditions. For any simple n-dimensional polyhedron:

(1) the H-polynomial is reflexive, i.e., $H_k = H_{n-k}$;
(2) coefficients of an H-polynomial increase to the middle, i.e., $H_0 \leq H_1 \leq \cdots \leq H_{[n/2]}$, and the coefficients H_0 and H_n are 1;
(3) for any i such that $1 \leq i < [n/2]$, the difference $H_{i+1} - H_i$ satisfies an estimate that depends on the difference $H_i - H_{i-1}$; more precisely, $H_{i+1} - H_i \leq Q^i(H_i - H_{i-1})$. Here Q^i is a special function of integers (for the definition of the function Q^i, see [14]). We shall need this function in subsection 2.3.

The statement about the reflexivity of the H-polynomial is also known as the Dehn-Sommerville Theorem. A special case of that theorem for $k = 0$ coincides with the Euler Theorem for polyhedra: $\sum_{i=0}^{n}(-1)^i F_i = 1$. Indeed, the number H_0 is equal to $\sum_{i=0}^{n}(-1)^i F_i = 1$, while H_n is equal to F_n, which is equal to one since an n-dimensional polyhedron has exactly one face in the highest dimension. The Dehn-Sommerville Theorem shows that the numbers F_i for a simple n-dimensional polyhedron obey $[(n+1)/2]$ linear relations. In particular, if we consider simple 3-dimensional polyhedra, we have, in addition to the Euler relation, the obvious relation $2E = 3V$ (where E is the number of edges and V the number of vertices).

The increase of the first half of the coefficients of an H-polynomial is far from trivial; there is still no elementary proof of this fact. The only known proof is based on the theory of toric varieties and algebraic geometry. We shall discuss this proof below. The increase of the first half of the coefficients of an H-polynomial implies the following

COROLLARY. *All coefficients $H_0, \ldots, H_n$ of an H-polynomial are positive.*

We are now going to prove this corollary and the Dehn-Sommerville Theorem using an elementary argument—a remarkable analogue of Morse theory in linear programming.

Let us say that a linear function on a simple polyhedron is *generic* if its restriction to any edge is not constant. We say that a generic linear function on a simple polyhedron has index i at a vertex if it decreases along exactly i edges emanating from the vertex (and increases along the remaining $n - i$ edges).

THEOREM [17]. *The number $h(i)$ of vertices of index i for any generic linear function on a simple polyhedron is exactly H_i, the ith coefficient of the H-polynomial of the polyhedron.*

This theorem generalizes the obvious fact that a generic linear function on a simple polyhedron (regardless of the choice of the function) has exactly one maximum (i.e., $h_n = 1$) and one minimum (i.e., $h_0 = 1$). Let us proceed to the proof.

It will suffice to show that if F_k is the number of k-dimensional faces of a simple polyhedron, then $F_k = \sum \binom{i}{k} h(i)$, $k = 0, \ldots, n$. Consider the map that assigns to each k-dimensional face of the polyhedron the vertex at which the linear function assumes its maximum on that face. Under this map, any vertex of index i will be assigned to exactly $\binom{i}{k}$ k-dimensional faces. Indeed, for any point on a k-dimensional face at which the function has a maximum, it decreases along the k edges from the point that lie on the face. Conversely, to any set of k edges from the same vertex, along which the function decreases, there corresponds a k-dimensional face, namely the face for which this vertex is a maximum point; in a simple polyhedron, every set of k edges radiating from one vertex spans a k-dimensional face. Now summing the number of inverse images of our mapping over all vertices, we obtain the required formula.

COROLLARY 1. $H_i = H_{n-i}$.

Indeed, a vertex of index i for a function L has index $n - i$ for the function $-L$. Hence the number $h(i)$ for the function L is precisely the number $h(n - i)$ for the function $-L$.

COROLLARY 2. *For a simple n-dimensional polyhedron the numbers* H_0, $H_1, \ldots, H_n$ *are strictly positive.*

Indeed, for any vertex of a polyhedron and any number $i = 0, 1, \ldots, n$, there exists a generic linear function that has index i at that vertex.

2.3. Simple polyhedra and quasismooth toric varieties. As already stated, for every Newton polyhedron one can associate a toric variety. If the Newton polyhedron is simple, then the variety is quasismooth, i.e., it is isomorphic near each point to a domain in the space $\mathbb{C}^n$ modulo the action of a finite group. The cohomology theory of quasismooth varieties is quite similar to the cohomology theory of smooth varieties. On quasismooth varieties one can consider differential forms—in the local charts these are forms in domains of $\mathbb{C}^n$ that are invariant under the action of the appropriate groups. On quasismooth varieties, just as on smooth ones, the cohomology carries a (pure) Hodge structure, and one can define numbers $h^{p,q}$ as dimensions of the Hodge subspaces in the $(p+q)$-dimensional cohomology. Just as for smooth varieties, there is the Poincaré duality for quasismooth varieties and we have equalities $h^{p,q} = h^{n-p,n-q}$, $h^{p,q} = h^{q,p}$. In particular, $h^{i,i} = h^{n-i,n-i}$. In addition, the Strong Lefschetz Theorem holds for such varieties. In particular, this theorem states that, for n-dimensional quasismooth varieties, $h^{0,0} \leq h^{1,1} \leq \cdots \leq h^{k,k}$, where $k = [n/2]$.

We have the following theorem: the number $h^{i,i}$ for the toric variety associated with a simple Newton polyhedron Δ coincides with the ith coefficient H_i of the H-polynomial of the polyhedron; all the other numbers

$h^{p,q}$, $p \neq q$, are equal to zero. It is now evident that the symmetry of the coefficients of the H-polynomial of a simple polyhedron is a consequence of Poincaré duality, while the increase of the first half of the coefficients of the H-polynomial of a simple polyhedron is a corollary of the Strong Lefschetz Theorem. In addition, for toric varieties the cohomology classes of Hodge degree (i, i) in the cohomology ring are generated by the cohomology classes of Hodge degree $(1, 1)$. In the geometry of polyhedra this is due to the fact that every proper face of a polyhedron is an intersection of proper faces of higher dimensions. Faces of dimension $n - i$ correspond to algebraic subvarieties of the toric variety of complex dimension $n - i$. The group of cycles spanned by these subvarieties is dual to the Hodge subspace in the cohomology of degree (i, i), and an intersection of faces corresponds to intersection of cycles and multiplication of dual cohomology classes.

For any n-dimensional quasismooth variety, whose cohomology of Hodge degree (i, i) is generated by its cohomology of Hodge degree $(1, 1)$, we have the inequality $(h^{i+1, i+1} - h^{i,i}) \leq Q^i(h^{i,i} - h^{i-1, i-1})$, where Q^i is the function appearing in McMullen's conditions and $0 < i < [n/2]$. Thus, the restriction on the rate of increase of the numbers H_i for simple polyhedra is also a consequence of toroidal geometry.

The linear programming theorem proved above was also suggested by toroidal geometry. The point is that there exists a special mapping onto a Newtonian polyhedron, called the *moment map* of the toric variety associated with the polyhedron [17]. Under this map, orbits of complex dimension i go into i-dimensional faces of the polyhedron. The superposition of a generic linear function on a Newton polyhedron and the moment map determines a Morse function on the toric variety. The critical points of this function are 0-dimensional orbits. The Morse index of a critical point is twice the index of the linear function at the corresponding vertex. Indeed, the function decreases on a $2i$-dimensional real variety containing the critical point and increases on a $2(n - i)$-dimensional real variety containing it. Here i is the index of the vertex, and the varieties of decrease and increase are the closures of the orbits corresponding to the i-dimensional and $(n - i)$-dimensional faces on which the function decreases and increases. It is obvious that if a Morse function on a variety has only critical points of even index, then the number of points of a fixed index $2i$ does not depend on the choice of the Morse function and is equal to the $2i$-dimensional Betti number of the variety. That is why toric geometry explains the linear programming theorem proved above. In addition, the above proof of the Dehn-Sommerville Theorem imitates the proof of Poincaré duality, which is based on Morse theory and compares the cell complexes of the variety corresponding to Morse functions f and $-f$.

2.4. Combinatorics of polyhedra and discrete groups in Lobachevsky spaces.
We now return to the geometry of simple polyhedra.

NIKULIN'S THEOREM [18]. *The average number of l-dimensional faces on a k-dimensional face of a simple n-dimensional polyhedron, $0 \leq l < k \leq (n+1)/2$, is bounded from above by a certain function of l, k, and n, which can be written down explicitly. If n tends to infinity, the function tends to the number of l-dimensional faces of a k-dimensional cube.*

Using Nikulin's Theorem, Vinberg [19] showed that in the Lobachevsky space of dimension ≥ 32 there are no discrete groups generated by reflections with a compact fundamental polyhedron. It is easy to see that the fundamental polyhedron of such a group is simple—this is why Nikulin's Theorem is applicable.

Nikulin proved his theorem using only the Dehn-Sommerville Theorem and the fact that the coefficients of the H-polynomial are positive. We have presented above very simple proofs of these properties based on linear programming. These simple proofs have enabled us to prove the following generalization of Nikulin's Theorem.

THEOREM [17]. *The bound in Nikulin's Theorem is valid not only for simple polyhedra, but also for edge simple polyhedra. (An n-dimensional polyhedron is said to be edge simple if each of its edges meets exactly $(n-1)$ faces of higher dimension.)*

Using this bound, we can prove the following

THEOREM [17, 20]. *In a Lobachevsky space of dimension > 995 there are no discrete groups generated by reflections with a fundamental polyhedron of finite volume.*

A fundamental polyhedron of finite volume is edge simple. That is just why the above bound is applicable. The bound on the mean number of l-dimensional faces on a k-dimensional face of an edge simple polyhedron has recently found an unexpected application in the algebraic geometry of surfaces.

2.5. Sections of simple polyhedra. To end this paper, we shall discuss yet another theorem whose proof is based on programming. It estimates the number of faces in any dimension of a generic l-dimensional section of a simple n-dimensional polyhedron with fixed numbers $F_0, \ldots, F_n$ of faces of different dimensions. Consider a simple $(n-l)$-dimensional polyhedron, the number f_k of whose k-dimensional faces, $k \geq (n-l)/2$, is equal to the number of $(k+l)$-dimensional faces of the initial polyhedron, $f_k = F_{k+l}$, and the number of faces of lower dimensions is found by the Dehn-Sommerville Theorem.

THEOREM [14]. *The number of faces in any dimension of a generic l-dimensional section does not exceed the number of faces in the same dimension of the polyhedron defined above.*

Of course, the theorem is obvious for faces of dimensions $\geq (n-l)/2$. For faces of lower dimensions, however, it gives a nontrivial bound.

EXAMPLES. 1. A plane that does not pass through the vertices of a simple 3-dimensional polyhedron with E edges can cut at most $E/3+2$ edges. This bound is sharp. By tilting the middle section of a prism with $3k$ edges we can make it cut the upper and lower bases, while still cutting all the lateral faces. Then the resulting section will cut exactly $k+2$ edges.

2. Consider a polyhedron with the combinatorial structure of an n-dimensional cube. If we let $n \to \infty$, a hypersurface can asymptotically cut at most a fraction $(2\pi)^{-1/2} \int_{-c}^{+c} \exp(-x^2/2)dx$ of the k-dimensional edges, where $k \sim c\sqrt{n}$. This bound is sharp.

3. What is the maximum possible number of k-dimensional faces in a simple l-dimensional polyhedron which has p faces of the highest dimension? The answer to this question is given by the famous Upper-Bound Theorem, proved in 1970 by McMullen [21]. Every simple l-dimensional polyhedron with p faces of the highest dimension can be considered as a generic l-dimensional section of a $(p-1)$-dimensional simplex. Application of our theorem in this case yields the Upper-Bound Theorem.

REFERENCES

1. A. G. Khovanskiĭ, *Newton polyhedra and the genus of full intersections*, Funktsional. Anal. i Prilozhen. **12** (1978), no. 1, 51–61; English transl. in Functional Anal. Appl. **12** (1978), no. 1.
2. A. G. Kushnirenko, *The Newton polyhedron and the number of solutions of a system of k equations in k unknowns*, Uspekhi Mat. Nauk **30** (1975), no. 2, 266–267. (Russian)
3. D. N. Bernstein, *The number of roots of a system of equations*, Funktsional. Anal. i Prilozhen. **9** (1975), no. 3, 1–4; English transl. in Functional Anal. Appl. **9** (1975), no. 3.
4. V. I. Danilov and A. G. Khovanskiĭ, *Newton polyhedra and algorithms for computing Hodge-Deligne numbers*, Izv. Akad. Nauk SSSR Ser. Mat. **50** (1986), no. 5, 925–945; English transl. in Math. USSR-Izv. **29** (1987), no. 2.
5. A. D. Aleksandrov, *Towards a theory of mixed volumes of convex bodies.* I, Mat. Sb. **2** (1937), no. 5, 947–972 (Russian); II, **2** (1937), no. 6, 1205–1238; III, **3** (1938), no. 1, 27–46; IV, **3** (1938), no. 2, 227–251.
6. B. Teissier, *Du théorème de l'index de Hodge aux inégalités isopérimétriques*, C. R. Acad. Sci. Paris Sér. AB **288** (1979), no. 4, A287–A289.
7. _____, *Variétés toriques et polytopes*, Sèm. Bourbaki, 1980/81, no. 555, Lecture Notes in Math., vol. 901, Springer-Verlag, Berlin and New York, 1981, pp. 71–84.
8. A. G. Khovanskiĭ, *Geometry of convex polyhedra and algebraic geometry*, Uspekhi Mat. Nauk **34** (1979), no. 4, 160–161. (Russian)
9. _____, *Algebra and mixed volumes*, A Series of Comprehensive Studies in Mathematics. Geometry (Y. D. Burago and V. A. Zalgaller, eds.), vol. 285, Springer-Verlag, Berlin and New York, pp. 182–207.
10. D. Mumford, *Lectures on curves on an algebraic surface*, Princeton Univ. Press, Princeton, N. J., 1966.
11. F. Minding, *Über die Bestimmung des Grades ein durch Elimination hervorgehenden Gleichung*, J. Reine Angew. Math. **22** (1841), 178–183; J. Math. Pures Appl., Ser. 1 **6** (1841), 412–418.
12. A. G. Khovanskiĭ, *Newton polyhedra and toral varieties*, Funktsional. Anal. i Prilozhen. **11** (1977), no. 4, 56–64; English transl. in Functional Anal. Appl. **11** (1977), no. 4.
13. G. Kempf, F. Knudsen, D. Mumford, and B. Saint-Donat, *Toroidal embeddings.* I, Lecture Notes in Math., vol. 339, Springer-Verlag, Berlin and New York, 1973.
14. P. McMullen, *The number of faces of simplicial polytopes*, Israel J. Math. **9** (1971), 559–570.

15. R. Stanley, *The number of faces of a simplicial convex polytope*, Adv. in Math. **35** (1980), no. 3, 236–238.
16. L. J. Billera and C. W. Lee, *Sufficiency of McMullen's conditions for f-vectors of simplicial polytopes*, Bull. Amer. Math. Soc. **2** (1980), 181–185.
17. A. G. Khovanskiĭ, *Hyperplane sections of polyhedra, toric varieties and discrete groups in Lobachevsky space*, Funktsional. Anal. i Prilozhen. **20** (1986), no. 1, 50–61; English transl. in Functional Anal. Appl. **20** (1986), no. 1.
18. V. V. Nikulin, *On arithmetical groups generated by reflections in Lobachevsky spaces*, Izv. Akad. Nauk SSSR Ser. Mat. **44** (1980), no. 5, 637–649, 719–720; English transl. in Math. USSR-Izv. **16** (1981), no. 3.
19. E. B. Vinberg, *The absence of crystallographic groups of reflections in Lobachevsky spaces of large dimension*, Funktsional. Anal. i Prilozhen. **15** (1981), no. 2, 67–68; English transl. in Functional Anal. Appl. **15** (1981), no. 2.
20. M. N. Prokhorov, *The absence of discrete groups of reflections with a noncompact fundamental polyhedron of finite volume in Lobachevsky spaces of large dimension*, Izv. Akad. Nauk SSSR Ser. Mat. **50** (1986), no. 2, 320–332; English transl. in Math. USSR-Izv. **28** (1987), no. 2.
21. P. McMullen, *The maximum number of faces of convex polytopes*, Mathematika **17** (1970), 179–184.

Translated by A. BOCHMAN

Recent Titles in This Series

(Continued from the front of this publication)

114 **M. Š. Birman and M. Z. Solomjak,** Quantitative Analysis in Sobolev Imbedding Theorems and Applications to Spectral Theory
113 **A. F. Lavrik,** Twelve Papers in Logic and Algebra
112 **D. A. Gudkov and G. A. Utkin,** Nine Papers on Hilbert's 16th Problem
111 **V. M. Adamjan, et al.,** Nine Papers on Analysis
110 **M. S. Budjanu, et al.,** Nine Papers on Analysis
109 **D. V. Anosov, et al.,** Twenty Lectures Delivered at the International Congress of Mathematicians in Vancouver, 1974
108 **Ja. L. Geronimus and Gábor Szegő,** Two Papers on Special Functions
107 **A. P. Mišina and L. A. Skornjakov,** Abelian Groups and Modules
106 **M. Ja. Antonovskiĭ, V. G. Boltjanskiĭ, and T. A. Sarymsakov,** Topological Semifields and Their Applications to General Topology
105 **R. A. Aleksandrjan, et al.,** Partial Differential Equations, Proceedings of a Symposium Dedicated to Academician S. L. Sobolev
104 **L. V. Ahlfors, et al.,** Some Problems on Mathematics and Mechanics, On the Occasion of the Seventieth Birthday of Academician M. A. Lavrent'ev
103 **M. S. Brodskiĭ, et al.,** Nine Papers in Analysis
102 **M. S. Budjanu, et al.,** Ten Papers in Analysis
101 **B. M. Levitan, V. A. Marčenko, and B. L. Roždestvenskiĭ,** Six Papers in Analysis
100 **G. S. Ceĭtin, et al.,** Fourteen Papers on Logic, Geometry, Topology and Algebra
99 **G. S. Ceĭtin, et al.,** Five Papers on Logic and Foundations
98 **G. S. Ceĭtin, et al.,** Five Papers on Logic and Foundations
97 **B. M. Budak, et al.,** Eleven Papers on Logic, Algebra, Analysis and Topology
96 **N. D. Filippov, et al.,** Ten Papers on Algebra and Functional Analysis
95 **V. M. Adamjan, et al.,** Eleven Papers in Analysis
94 **V. A. Baranskiĭ, et al.,** Sixteen Papers on Logic and Algebra
93 **Ju. M. Berezanskiĭ, et al.,** Nine Papers on Functional Analysis
92 **A. M. Ančikov, et al.,** Seventeen Papers on Topology and Differential Geometry
91 **L. I. Barklon, et al.,** Eighteen Papers on Analysis and Quantum Mechanics
90 **Z. S. Agranovič, et al.,** Thirteen Papers on Functional Analysis
89 **V. M. Alekseev, et al.,** Thirteen Papers on Differential Equations
88 **I. I. Eremin, et al.,** Twelve Papers on Real and Complex Function Theory
87 **M. A. Aĭzerman, et al.,** Sixteen Papers on Differential and Difference Equations, Functional Analysis, Games and Control
86 **N. I. Ahiezer, et al.,** Fifteen Papers on Real and Complex Functions, Series, Differential and Integral Equations
85 **V. T. Fomenko, et al.,** Twelve Papers on Functional Analysis and Geometry
84 **S. N. Černikov, et al.,** Twelve Papers on Algebra, Algebraic Geometry and Topology
83 **I. S. Aršon, et al.,** Eighteen Papers on Logic and Theory of Functions
82 **A. P. Birjukov, et al.,** Sixteen Papers on Number Theory and Algebra
81 **K. K. Golovkin, V. P. Il'in, and V. A. Solonnikov,** Four Papers on Functions of Real Variables
80 **V. S. Azarin, et al.,** Thirteen Papers on Functions of Real and Complex Variables
79 **V. I. Arnol'd, et al.,** Thirteen Papers on Functional Analysis and Differential Equations
78 **A. V. Arhangel'skiĭ, et al.,** Eleven Papers on Topology
77 **L. A. Balašov, et al.,** Fourteen Papers on Series and Approximation

(See the AMS catalog for earlier titles)